PHYSICAL GEOGRAPHY

THE BASICS

Joseph Holden

Routledge
Taylor & Francis Group

LONDON AND NEW YORK

First published 2011
by Routledge
2 Park Square, Milton Park, Abingdon, Oxon, OX14 4RN

Simultaneously published in the USA and Canada
by Routledge
711 Third Avenue, New York, NY 10017

Routledge is an imprint of the Taylor & Francis Group, an informa business

British Library Cataloguing in Publication Data
A catalogue record for this book is available from the British Library

Library of Congress Cataloging in Publication Data
Holden, Joseph.
  Physical geography: the basics/Joseph Holden.
  p. cm.
  Includes index.
  1. Physical geology. I. Title.
  QE28.2.H635 2011
  551–dc22   2010046796

ISBN: 978-0-415-55929-4 (hbk)
ISBN: 978-0-415-55930-0 (pbk)
ISBN: 978-0-203-81714-8 (ebk)

Typeset in Bembo by
Wearset Ltd, Boldon, Tyne and Wear

MIX
Paper from
responsible sources
FSC
www.fsc.org FSC® C004839

Printed and bound in Great Britain by
TJ International Ltd, Padstow, Cornwall

PHYSICAL GEOGRAPHY
THE BASICS

volume provides an excellent and easily digestible introduction
physi a geography. I would recommend all prospective geography
le s to read this before going to university.

Dr Martin Evans, *University of Manchester*

his book offers the undergraduate student an introduction to the
topics that will form the foundation of any Geography course, and will
continue to be a vital reference book throughout their studies. It is an
excellent read!

Dr Simon Jones, *University of Wales*

Physical Geography: The Basics is a concise and engaging introduction to the
inte s, systems and processes that have shaped, and continue to shape,
the al world around us. This book introduces five key aspects of the
st of physical geography:

* mosphere, weather and climate systems
* the carbon cycle and historic and contemporary climate change
* plate tectonics, weathering, erosion and soils
* the role of water and ice in shaping the landscape and impacting human
  activity
* the patterns of plant and animal life and human impacts upon them.

The book features diagrams, maps and a glossary to aid understanding of key
ide and suggestions for further reading to allow readers to develop their
in est in the subject – making *Physical Geography: The Basics* the ideal start-
in point for anyone new to the study of geography and the environment.

j **Holden** is Chair of Physical Geography at the University of Leeds,
UK e is the editor and contributor of three chapters of the market-leading
textbook A nd author
of over 10

The Basics

CONTENTS

ILLUSTRATIONS

Figures

Boxes

ACKNOWLEDGEMENTS

This book brings together wide-ranging subject areas from climate change, carbon and weather to oceans, mountain building, volcanoes, soils, hydrology, glaciers and ecosystems. The book covers an extremely broad set of topics for each of which there could easily be a separate book in the 'Basics' series. I hope that I have met the challenge and produced an informative book providing a basic level of understanding and terminology for a range of topics under the umbrella of 'physical geography'. This book could only be produced because of the dedicated work of the thousands of scientists who have tried to understand the way the world works. I acknowledge these contributions to our understanding, both large and small, by researchers from around the world as I attempt to synthesise this understanding in this book. I apologise to these scientists for any oversimplification or misunderstanding that might be evident from my description of processes or features of the Earth's physical geography.

I would like to thank my family for coping with my long hours while I wrote this book. The support of my wife Eve and her strength in looking after our three young daughters, Justina, Mary and Alice was an inspiration. I also thank Elizabeth Pettifer for helping to organise the glossary and index and Alison Manson for drawing up the figures.

Figure 2.3 was kindly reproduced from Figure SPM.3, page 6 of the IPCC (2007) report as follows: Summary for Policymakers, in *Climate Change 2007: The Physical Science Basis*, Working Group I Contribution to the Fourth Assessment Report of the Intergovernmental Panel on Climate Change, Solomon, S., Qin, D., Manning, M., Chen, Z., Marquis, M., Averyt, K.B., Tignor, M. and Miller, H.L. (eds), Cambridge University Press.

INTRODUCTION

Physical geography is the study of the interactions between the Earth's climate system, landscapes, oceans, plants, animals and people. Physical geography is therefore a highly relevant and topical subject for society and the processes studied affect our daily lives. Ways of tackling many of the world's challenges are underpinned by an understanding of physical geography. For example, every time there is a major disaster such as a wildfire, earthquake, volcanic eruption, flood, drought, landslide or tropical storm, these can only be understood through integrated understanding of land–ocean–atmosphere–human interactions. Sometimes what seems to be a natural disaster (e.g. a major flood event) has been impacted by human activity (e.g. deforestation of hillsides in the upper parts of river basins), making the event worse. At the same time, humans have a creative ability to come up with solutions to major problems or at least make things more resilient to disasters (e.g. buildings with engineering design to allow them to stay standing during an earthquake). Solutions work best when the way the Earth's system works is fully understood. To gain this understanding is the business of physical geography.

Knowing about climate change and its impacts is necessary so that we can adapt to changes and mitigate effects to minimise the damage to human life and infrastructure. Knowing about climate, soil, plant and water interactions is crucial to supplying food and clean water as the world's population grows from 6 billion to 9 billion over the next 40 years. Therefore, understanding the basics of physical geography should be a key area of knowledge owned by the people of planet Earth and should be a fundamental area of understanding for the policymaking community.

This book attempts to impart a basic understanding of major topics covered by physical geography. Material has been grouped into five chapters. The first two chapters cover the atmosphere, weather and climate system and the history of global environmental change, contemporary climate change and carbon. The next two chapters deal with processes that shape the Earth's physical appearance and movements of water and sediment while the final chapter deals with **biogeography** [bold words in the main text are glossary terms], which examines processes responsible for the spatial distribution of plants and animals. Because the Earth's system is all about interactions then there are inevitable overlaps between chapters. I have tried to write this book in an accessible manner providing a basic overview of the subject including examples and boxed features to illustrate key points. I have not tried to describe every environment on the planet, but instead tried to impart an understanding of processes that would then enable the reader to explain why and how an environment on the planet functions in a particular way. The further reading lists at the end of each chapter can be explored to develop understanding of processes in more detail for those areas that interest you most.

ATMOSPHERE, OCEANS, WEATHER AND CLIMATE

COMPOSITION OF THE ATMOSPHERE

The layer of gases surrounding the Earth rises about 500 kilometres above the surface, although there is no distinct boundary between the Earth's atmosphere and space. However, three-quarters of the gas in the atmosphere is within 11 kilometres of the Earth's surface as the density of gas is very small at high altitudes. This is why it becomes more difficult to breathe when you climb mountains and why people need oxygen masks if they are in the open air at high altitudes. The lower layer of the atmosphere is called the **troposphere** and extends to about 6 kilometres altitude over the poles and about 15 kilometres over equatorial regions. This layer is the one in which air mixes most rapidly and where we experience weather.

The Earth's atmosphere acts as a filter protecting us from space debris and harmful radiation. The Earth receives only two-billionths of the Sun's total energy release but this energy is the main driver for water, air and wind motions and most life on Earth. It is therefore important to understand how the energy from the Sun drives these processes. When the Sun's energy reaches the Earth about 6 per cent of it is scattered and returned to space by the atmosphere, 21 per cent is scattered and reflected by clouds and 18 per cent is absorbed by the atmosphere and clouds temporarily before being sent back out to space. Of the 55 per cent that reaches the Earth's surface 4 parts are reflected back to space by reflective surfaces such as ice sheets, snow and dry, light, sandy soils, while 51 parts are absorbed by the surface. This means that of the total solar energy received by the Earth, only around a half makes it all the way down to warm the Earth's surface. Some of this energy is used for processes such as evaporation of water or for plant growth. However, most absorbed radiation from the Sun (known as short-wave radiation; an example of short-wave radiation is visible light) is transformed by the land, oceans and vegetation and emitted back into the atmosphere as long-wave radiation in the form of heat energy (invisible **infrared radiation**). Except for the 19 per cent of incoming solar energy that is temporarily absorbed, the atmosphere is mostly transparent to incoming short-wave radiation. This means, perhaps surprisingly to many people, that the air is mainly heated from below by long-wave heat energy emitted by the Earth's surface. Thus, the atmosphere should be warmer close to the Earth's surface but cool with altitude in the troposphere. Since the air is warmed by the surface below, this means that during the day the air near the surface becomes less dense and more buoyant. Less dense gases or liquids will naturally seek to rise and more dense fluids will seek to fall. Hence the less dense air near the surface seeks to rise above cooler, denser air which in turn sinks towards the Earth's surface. As the air rises it in turn cools because it is able to expand due to the lower air pressure at higher altitudes. The reason it cools is due to a fundamental law of nature which means that as the pressure of a gas decreases the temperature will decrease. The result of these processes is that there is large scale vertical mixing of the air within the troposphere as rising warm air is replaced by cooler descending air.

The atmosphere is made up of mainly nitrogen (78 per cent) and oxygen (21 per cent). The remaining 1 per cent is made up of mainly argon. There are also small concentrations of other gases such as hydrogen, water vapour (the gaseous form of water), methane, nitrous oxide, ozone and carbon dioxide. However, despite their low concentrations some of these other gases are important for the climate we experience. While the gases of the atmosphere are almost unaffected by the short-wave radiation provided by the Sun, some of them readily absorb long-wave radiation produced by the Earth's surface. Unlike oxygen and nitrogen, some gases such as carbon dioxide, methane, water vapour and nitrous oxide absorb the thermal energy emitted by the Earth's surface and provide a sort of blanket over the Earth. They radiate this energy back down to Earth again which in turn is absorbed by the Earth and this further enhances the heating of the atmosphere. A greenhouse does a similar thing. The glass allows short-wave radiation to pass through it to the soil and plants which then absorb the radiation and re-radiate thermal energy back towards the glass. However, the glass traps the long wavelength heat energy and the warmer air inside the greenhouse. The **natural greenhouse effect** in the atmosphere is a good thing. If this did not happen then during the day the Sun's energy would be absorbed by the land, oceans, and vegetation at the surface and then transformed into heat which would be radiated back into the atmosphere. However, at night all of this energy would radiate back into space and so the Earth's surface temperature would fall to extremely cold levels very quickly. The greenhouse gases prevent this from happening by retaining some of the energy within the troposphere, delaying its release back out to space, and keeping the planet at a good temperature for life. The average temperature of the Earth's surface is 15°C but without the natural greenhouse effect the average temperature across the Earth would be around −20°C.

The composition of the atmosphere has changed through time. The Earth is around 4.6 billion years old. The early atmosphere mainly consisted of nitrogen gas and carbon dioxide with no oxygen gas. Oxygen gas did not start to appear in the atmosphere until about 2 billion years ago. It was at this point that bacteria evolved and they functioned by absorbing carbon dioxide from the atmosphere and then releasing oxygen through **photosynthesis**. More recently humans have also changed the composition of the atmosphere slightly through the burning of fossil fuels and release of other chemicals. This topic is explored further in Chapter 2.

There are many other complex feedbacks between the atmosphere and the Earth which control climate. We will turn to some of these in later chapters but for now a good example is the growth of peatlands. Peat is an extremely carbon-rich soil composed of dead plant matter that has not fully decayed and which builds up on the land year after year. In some places it has built up deposits over 10 metres thick. It forms where the conditions are waterlogged because waterlogging slows the rate of decay when plants die. Plants take carbon out of the atmosphere to form their structure by incorporating it into carbohydrates. Once the plants die, if this carbon is not released again by decay, then it remains stored on land. In fact peatlands have preserved many interesting archaeological features including almost perfectly preserved prehistoric human remains with their leather shoes still intact. The world's peatlands are a large store of carbon that was once in the atmosphere, and they have actually reduced the amount of the greenhouse gas, carbon dioxide. Peatlands have helped to cool the climate by a few degrees. However, this all means that these peatlands could also be a large potential source of carbon dioxide for the atmosphere if they are rapidly degraded by human action such as through drainage or extraction for horticulture or fuel.

LARGE SCALE ATMOSPHERIC CIRCULATION

The atmosphere is in constant motion. Air moves vertically, primarily due to heating from the Earth's surface below. This motion of fluids (air is a fluid) is called **convection** and is the dominant process for transferring heat upwards from the Earth's surface. Because air pressure falls with height, then rising air expands and therefore cools. The rate of temperature change with altitude is known as the **lapse rate**. The normal rate of temperature change with altitude is known as the **environmental lapse rate** and is around 6.4°C per kilometre, but this is variable. However, a rapid temperature change associated with a rising and expanding parcel of air is described as 'adiabatic', meaning that there is no interchange of heat between the rising air parcel and its surroundings; the temperature change is internal to the air parcel. The rate of temperature decrease with altitude in the rising air parcel under these conditions is known as the **dry adiabatic lapse rate** and is 9.8°C per kilometre. Cooling of a rising air parcel may result in the air becoming saturated with water vapour, condensation of water droplets and the formation of clouds and **precipitation**. When water vapour condenses into liquid water it releases heat which then warms the air slightly and therefore the lapse rate within this air parcel is less than the dry adiabatic lapse rate and is known as the **saturated adiabatic lapse rate**. The exact value of this lapse rate varies depending on the amount of moisture in the air and the temperature. When some air rises, other air descends to replace it and so the troposphere is continually mixing.

As long as no condensation of water vapour occurs, a rising air parcel will cool at the dry adiabatic lapse rate of 9.8°C per kilometre, while the surrounding air changes temperature at the lower rate of the environmental lapse rate. However, condensation of water vapour may occur. Warmer air can hold more water vapour than cooler air. When an air mass holding water vapour has cooled sufficiently so that it is saturated (the **dew point**), condensation of water occurs. (Think of a cold drink you take out of the fridge and place in a warm room; water droplets form on the outside of the drink container as the air around it has cooled to the dew point and becomes fully saturated leading to condensation of water vapour onto the container.) The condensation process releases heat which warms the air and counteracts the cooling that results from expansion of the rising air. This warming may force the air even higher to form large clouds (the ones we see in the sky) of condensed water. So it is the difference between the lapse rates that determine whether there will be a continual rise of the air mass and cloud formation, or whether there will be very stable conditions (i.e. when the environmental lapse rate is less than the dry and saturated adiabatic lapse rates).

The above processes highlight the important role of water vapour in atmospheric motion. On average the atmosphere holds about a fortieth of global annual rainfall (about 25 millimetres depth of water if it were all deposited evenly across the whole Earth). Regular evaporation from land and oceans maintains rainfall throughout the year in different parts of the world. This is not evenly spread across the planet and some areas provide a lot more evaporation than they receive as rainfall, and vice versa, indicating that water moves significant distances in the atmosphere. As we will see later on, evaporation is also important for controlling major ocean circulations which have a big impact on the Earth's climate.

The above description of vertical atmospheric processes does not explain why winds and water vapour move horizontally across the Earth or why the same location can experience periods of calm and periods of storminess. The Earth's wind circulation patterns help form the climatic zones. There are two main processes that drive global wind circulation. The first is the uneven distribution of the Sun's radiation over the Earth's surface due to its spherical shape. Figure 1.1 shows how the same amount of solar radiation is spread over a large area near the poles compared to the equatorial regions where it is more concentrated. This creates a north–south temperature gradient and as heat is always transferred from hot materials to cooler materials then the warm air (and oceanic water) from the equator will naturally try to rise and move polewards at high altitudes within the troposphere to be replaced by lower level winds/ ocean currents moving in the opposite direction.

The second driver of global wind circulation is the Earth's rotation. If the Earth did not rotate and only one side faced the Sun then surface winds would blow from the cold dark side to the hot daylight side as rising air over the hot side would need to be replaced by colder air drawn in. However, the Earth's rotation acts to create an apparent deflection of winds to the right in the northern hemisphere and to the left in the southern hemisphere which is a process known as the **Coriolis effect** and which is stronger as you move toward the poles.

The other effect of the Earth's rotation relates to the tilt of this rotation and its impact on the seasons. The rotation of the Earth is not perpendicular to its rotation around the Sun and is tilted to 23.5° as shown in Figure 1.1. This drives seasonal cycles upon

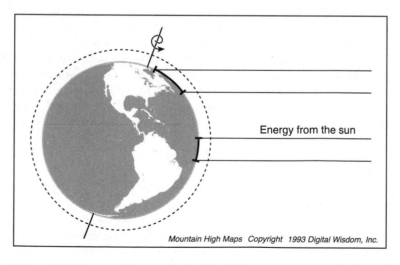

Figure 1.1 Diagram showing how the Sun's energy is more concentrated near the equator and more diffuse at the poles.

Earth because the Sun appears overhead at midday at the Tropic of Cancer (23°27'N) on 21–22 June and the Tropic of Capricorn (23°27'S) on 22–23 December. Areas that are polewards of the Arctic Circle (66°33'N) have at least one full 24 hour period of daylight on 21–22 June with the same being true for the Antarctic Circle on 22–23 December. There is almost six months of darkness in the winter at the poles and six months of summer daylight. Thus the northern hemisphere receives more of the Sun's energy from the March equinox (21/22 March) to the September equinox (21/22 September) than the southern hemisphere with the reverse true for the other half of the year. The midday Sun is directly overhead at the equator during equinoxes and on these days the length of day and night across the Earth are the same in all places.

The Coriolis effect combined with the latitudinal temperature gradients results in large atmospheric circulation cells as shown in Figure 1.2 with zones of rising and falling air creating low and high pressure at the surface. The Hadley cells are formed by rising air near the equator which flows towards the poles and then sinks at about 30° north and south before returning at low levels back to the equatorial regions. As air sinks (and thereby warms) creating high pressure at around 20° to 30° latitude (latitude refers to how far north or south of the equator you are with 0° at the equator and 90° at the poles), moisture condensation is not common and so this region consists of clear skies and light winds. It is at this high pressure zone where most of the world's deserts are to be found. Between 30° north and south there are equatorward-flowing easterly winds (i.e. winds moving from east to west) known as the **trade winds**. These meet from north and south near the equator at a zone of low pressure caused by rising warm air. This zone is known by atmospheric scientists as the **intertropical convergence zone**. Here conditions are favourable for warm moist rising air, condensation of water vapour, cloudiness and large amounts of rainfall. The intertropical convergence zone is only a few hundred kilometres wide.

Over the poles there are similar circulation cells to the Hadley cells but with descending cold air at the poles and air flowing towards the equator at the surface. However, the Coriolis effect diverts the surface polar winds in an easterly direction as shown by the arrows on the schematic diagram of the Earth in Figure 1.2.

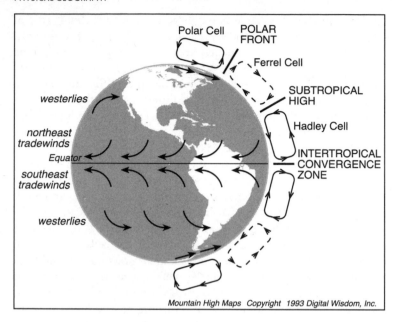

Mountain High Maps Copyright 1993 Digital Wisdom, Inc.

Figure 1.2 Schematic representation of the major atmospheric circulation cells and surface wind directions.

The polar cells tend to be weak as the Sun's energy is less intense here.

Between the Hadley cell and the polar cell is the Ferrel cell. The Ferrel cell consists of sinking air at around 30° north and south and poleward-moving westerly winds at the Earth's surface. Rising air occurs at the boundary between the Ferrel cell and the polar cell and this is called the **Polar front**. At the surface within the Ferrel cell are westerly winds carrying eastward moving cyclonic (anticlockwise) and anticyclonic (clockwise) circulation systems.

At higher levels in the troposphere the belt of prevailing westerly winds associated with the Ferrel cell is disturbed by large undulations. There are generally between three and six large atmospheric wave formations looping around the Earth known as **Rossby waves**. Imagine a dark spherical chocolate pudding on top of which you have just poured some white liquid icing. The icing will settle and solidify over the pudding to form a wavy boundary around the sides and this is how the Rossby waves would look if you could see them around the Earth. Their location is affected by large mountain ranges so that the waves, particularly in the northern hemisphere (where there is more land) are locked into preferred locations. In the winter there tends to be two troughs in the waves near the eastern edges of North America and Asia with ridges over the Pacific and Atlantic. This corresponds to troughs being associated with cold air over the winter continental land masses and ridges over the warmer oceans.

Within the Rossby waves are fast moving bands of air (at least 30 metres per second) called **jet streams** which are caused by sharp temperature gradients. Jet streams tend to be thousands of kilometres long, hundreds of kilometres wide and several kilometres deep. Aircraft can make use of the jet streams by travelling within them when moving west but avoiding them when moving east. For instance, the flight from the west coast of the USA to Europe can be an hour shorter than the reverse trip. The shape of the Rossby waves in the high level westerlies and the location of jet streams are used in weather forecasting as they are associated with the formation of large circulating air masses across the Earth's surface in the mid-latitudes. These circulating surface air patterns bring warm and cold air masses together and can lead to different calm/windy or dry/wet conditions.

LARGE SCALE OCEANIC CIRCULATION

The oceans cover 71 per cent of the Earth. The interactions between the oceans and the atmosphere play a major role in controlling the climate and weather at the Earth's surface. There are four primary factors that control ocean circulation:

(i) water **salinity**;
(ii) water temperature;
(iii) surface winds; and
(iv) the Coriolis effect.

(i) Contributing to water salinity, chemical weathering of rocks on land produces dissolved materials that enter the oceans. Some of these dissolved chemicals undergo reactions within the ocean to take them out of the water and deposit them on the ocean floor. For example, sea creatures extract dissolved calcium from the water to build their shells and when they die this calcium drifts onto the ocean floor. The balance of inputs from rivers and losses onto the ocean floor largely controls the chemical composition of the oceans. The concentration of salts in the ocean therefore varies depending on location and time. Surface water salinity can be diluted by melting ice or rainwater, or it can be concentrated by evaporation. The salinity of the ocean water around the dry subtropics of 20° to 30° north or south tends to be greater than elsewhere, because evaporation is greatest here.

(ii) In terms of water temperature, the surface of the ocean absorbs the Sun's energy and gains heat. More heat is gained than lost in the low latitudes and more is lost than gained in the high latitudes. Just as with the atmosphere there is a tendency for warm surface water to move polewards. This transport of heat from the equator towards higher latitudes provides good regulation of the Earth's climate system otherwise the poles would be even colder and the tropics even hotter than they are at present. Water provides an excellent long-term store of heat for the Earth, as it takes more energy to heat water by 1°C (and more energy is released when water cools by 1°C) than for any other substance.

Both temperature and salinity control the density of seawater. If the density of seawater increases with depth then the water is said to be vertically stable. If, however, there is more dense water on top of less dense water then vertical mixing of water will take place. This means that if there is a warm, strong wind encouraging evaporation at the ocean surface then this would lead to more salty and dense surface waters which leads to instability and mixing of water.

. (iii) The surface currents of the ocean as shown in Figure 1.3 are driven by the surface winds. For example, the trade winds (shown in Figure 1.2) drive the northern and southern equatorial currents moving in a westerly direction parallel to the equator.

(iv) The surface currents are deflected by the continents and the Coriolis effect (to the right in the northern hemisphere and to the left in the southern hemisphere). This creates warm currents along the eastern coasts of the Americas, Australia, Asia and Africa. In the North Atlantic this warm current is called the Gulf Stream which brings warm conditions to north-west Europe. The Gulf Stream forms the western and northern parts of the North Atlantic subtropical **gyre**. The five subtropical gyres (see Figure 1.3) are the

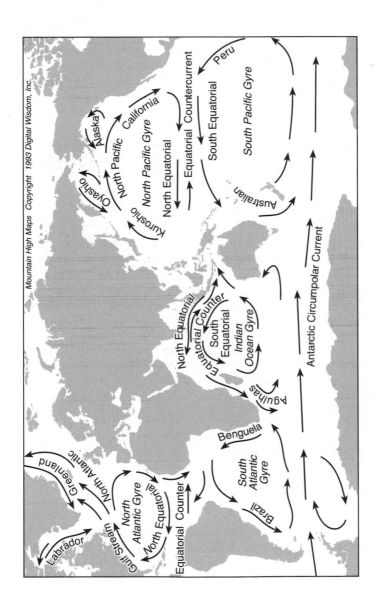

Figure 1.3 A map of the main surface ocean currents.

most dominant surface features of the world's oceans with their centres located at 30° north or south. The centres of the gyres were often avoided by sailors of the past due to their calm wind and calm ocean current conditions. Interestingly, the centres of these gyres have become collecting areas for the floating rubbish we have discarded into the world's oceans (Box 1.1). In each ocean there is a westerly current in the temperate mid-latitudes with a cold current flowing back towards the equator on the western edge of the continents. There is also a circumpolar current around Antarctica around 60° south. This is not mirrored around the Arctic as there are too many land masses in the region acting as a barrier to water flow.

Box 1.1 The oceanic floating plastic rubbish dump

The United Nations estimates that 10 per cent of the plastic we produce ends up being dumped in the oceans with around 18,000 pieces per square kilometre. The ocean currents move the floating debris of plastic bags, bottles, cartons, caps and so on which can later collect in the calm centres of the gyres. It is thought that in the centre of the North Pacific gyre the plastic is concentrated to around 334,000 pieces per square kilometre which may be equivalent to around 100,000 tonnes of plastic waste. Sea creatures can become entangled in the plastic. Plastic slowly breaks down into smaller pieces under sunlight, wave action and rubbing but does not fully degrade. This smaller plastic gets eaten by sea creatures and small fish may even mistake the particles for the tiny plants known as plankton that normally forms their diet. In some parts of the centre of the gyre there may be six times as many tonnes of plastic than plankton. The plastic accumulates harmful toxic chemicals which poisons sea creatures.

While surface currents are largely driven by wind, deep ocean currents are driven by differences in water density. This deep ocean circulation system is called the **thermohaline circulation** system (Figure 1.4). There are two important areas where deep water currents form. The first is in the North Atlantic/Arctic ocean and the second is in the Antarctic ocean. In the far North Atlantic, salty water from the Gulf Stream moves north into the Arctic and cools. Being saline and cool it is denser than surrounding waters and so it

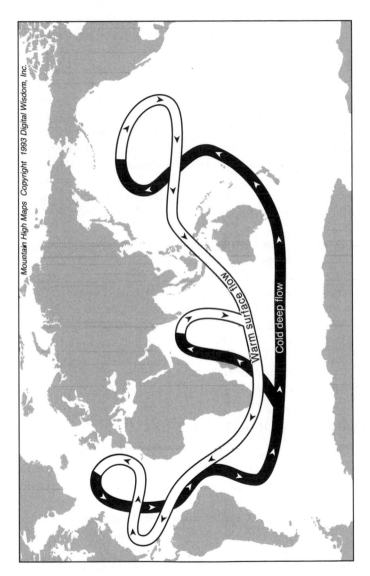

Warm surface flow

Cold deep flow

Figure 1.4 A schematic map of the thermohaline circulation system.

sinks and then flows south forming the main deep water current of the whole Atlantic. There have been concerns (most clearly exaggerated in the Hollywood movie *The Day After Tomorrow*) that ice melt in the North Atlantic region caused by global warming might supply lots of fresh water which would reduce the salinity of the Gulf Stream water so that it is not dense enough to sink. This would then cause the whole thermohaline circulation system to cease operating, or severely weaken, thereby reducing heat transfer from the equator resulting in much colder climate conditions at higher latitudes. For example, the current mild climate of north-west Europe could resemble the much colder climate of north-east Canada (e.g. Labrador). This would lead to further feedbacks as reflection of the Sun's energy by enhanced snow and ice cover in Europe would cool the climate further. There is some evidence that this slowing of the thermohaline circulation has occurred in the past and this is discussed in Chapter 2.

Sinking water has to be balanced by rising water coming to the surface. Deep water upwelling occurs thousands of kilometres away from the sinking zones. Along several eastern edges of continents where surface water is driven offshore by winds this surface water is replaced by deep water from below. This upwelled water is often rich in nutrients that have fallen to the depths of the ocean. When these nutrients reach the surface at upwelling zones, and where there is sufficient light, the nutrients can be utilised by plankton which in turn can maintain rich fisheries. One such example of a rich upwelling zone is off the coast of Peru.

INTERANNUAL CLIMATE VARIABILITY

While there are rich fisheries off the coast of Peru because of an important upwelling zone, these fisheries often collapse because of a specific event which directly demonstrates the feedbacks between the oceans and the atmosphere in controlling the climate. This event is known as the **El Niño Southern Oscillation** and this seems to occur about once every five years (but sometimes after three to seven years).

The Southern Oscillation is characterised by an exchange of air between the south-east Pacific (high pressure) and the Indonesian equatorial region (low pressure). Most of the time the trade winds are strong and converge over the warm waters of the western tropical Pacific where there is low pressure and lots of rainfall. During this time the ocean surface in the eastern tropical Pacific is relatively cold and the air above the ocean and the coastal parts of equatorial South America is cold and dry. However, every few years for 6 to 18 months the trade winds relax. As air pressure falls in the east the warmer surface ocean water and heavy rain moves east while there is a rise in surface air pressure in the west. This event is known as El Niño and brings drought to Indonesia and heavy rainfall and floods to coastal areas in equatorial South America. The movement of the warm water to the east means that the normal upwelling off Peru is capped with warm water preventing deep cold water from moving up to the surface. Thus, the nutrients that support the rich fisheries are cut off.

The effects of El Niño can also be seen across the planet due to the fact that the whole global climate system is linked and a change in one location has knock-on effects elsewhere. For example, contrary to normal conditions most El Niño winters are warm and dry over western Canada and wet from Texas to Florida. Damage from floods and landslides caused by very high rainfall in southern California has been linked to El Niño along with Indonesian forest fires, Australian bush fires and drought, and crop failures and famine in south-central Africa (e.g. Zambia, Zimbabwe, Mozambique and Botswana). Some El Niño events can be more intense and last longer than others. Forecasters have been trying to establish the likely nature and severity of each El Niño event in advance so that governments, farmers and other interested parties can prepare for the changes (e.g. by sowing different crops than normal or saving additional water before drought develops).

The El Niño Southern Oscillation is not the only example of interannual variability. For example, there is the North Atlantic Oscillation which when in a 'positive' phase increases the rainfall across northern Europe (with less across southern Europe) and results in milder northern European winters, while in a negative phase there will be less rainfall in northern Europe and more in southern Europe and north Africa. The North Atlantic Oscillation can remain in a positive or negative phase for several years or even decades.

It is also worth noting that natural events can cause variability in the Earth's climate. For example, volcanic ash from the eruption of

Mount Pinatubo in 1991 in the Philippines darkened skies around the world for over a year. The dust reflected more of the Sun's energy back into space and therefore the Earth was a little cooler that year. Unusually cold temperatures caused crop failures and famine in North America and Europe for two years, following the eruption of the volcano Tambora in 1815.

REGIONAL CLIMATE AND WEATHER

Climate is a long-term average of daily weather conditions occurring at a location. The decline in the Sun's energy received at the Earth's surface with latitude is important for determining the climate of a particular region. However, the distribution of oceans and continents and the circulation of the oceans and atmosphere are also important. This means that two locations at the same latitude can experience very different climates and different types of weather conditions. Northern Scotland has mild winters while Labrador, at the same latitude in north east Canada, has very cold winters.

Polar climate and weather

Polar climates consist of two main categories: ice cap and tundra. Polar ice caps are found in central Greenland and the Antarctic. They are dominated by high pressure and are extremely cold. Summer temperatures are generally below 0°C and winter temperatures below −40°C. In parts of the Antarctic the average annual temperature can be close to −50°C and temperatures close to −90°C have been recorded there. The air is very dry and there is little precipitation. Most of the polar ice caps can be officially classified as deserts as they have such little precipitation, typically less than 100 millimetres per year. This is why many physical geographers differentiate between hot deserts and cold deserts. The ice cap environment is made even worse for survival by the strong winds which occur because the ice caps cool the air around them causing a sinking of air from the high centre of the ice caps (the Antarctic interior is 3,500 metres high) to the coast.

Polar tundra climates are found in northern Scandinavia, Siberia, Iceland, coastal Greenland and high latitudes of North America (see Figure 5.1 in Chapter 5). The temperature of the warmest month in the polar tundra is above 0°C but below 10°C. Winter temperatures are generally low (average below –25°C). Mean annual precipitation tends to be less than 300 millimetres. The weather is dominated by the prolonged winter with dry, clear high pressure conditions. Although summer sunshine is weak in polar tundra regions, the long daylight hours can result in melting of the snow cover for short periods, allowing the upper soil layers to thaw. This provides a short growing season (see Chapter 5).

Mid-latitude climate and weather

The mid-latitudes are dominated by weather systems that move across the Earth. Air speeds up in high-level Rossby waves (I'm sure you remember the chocolate pudding icing example from earlier). This faster movement causes sections of air to spread apart (there is divergence of the air). Upper air convergence occurs where air slows down in the Rossby wave. If the divergence of air high up in the troposphere is greater than the convergence of air down below near the Earth's surface then there will be a fall in surface air pressure and air will rise. This, along with the Coriolis force, leads to the formation of a large, circulating, rising mass of air flowing in an anticlockwise direction. This is known as a depression. If the opposite occurs then there will be a zone of high pressure at the surface with descending air flowing in a clockwise direction. This is known as an anticyclone. Descending air in anti-cyclones results in the air warming which means that it is less satu-rated (less close to the dew-point temperature) and typically means clear conditions. However, on some occasions, there can be a **temperature inversion** whereby colder, moist air is trapped below the warmer air and a layer of cloud or fog can form, espe-cially in winter. During summer the air below the inversion is warmed sufficiently to cause the cloud or fog to dissipate.

A weather front is the boundary between two masses of air of dif-ferent densities (and temperature). A cold front is where a cold air mass is moving into a warm air mass whereas a warm front is the reverse. Box 1.2 outlines the characteristics of air masses. The rising air in depressions is concentrated along the warm and cold fronts. Rising air will produce condensation of water vapour leading to the formation of cloud and precipitation. Rising air in depressions takes place over a large area and while there may not always be heavy precipitation there can be substantial amounts of precipitation over a large area. Looking at vertical slices through fronts shows they tend to gently slope at a rate of 1 metre of vertical rise for every 80 to 150 metres of lateral distance. Cold fronts tend to be steeper than warm fronts and over time a cold front tends to overtake a warm front leading to an **occluded front**. As a cold front passes there can be a sudden drop in temperature experienced at ground level (5°C fall within 30 minutes is not uncommon).

Box 1.2 Air masses

An air mass is a regional parcel of air which has developed over an area where it has remained for a period of days and gained particular temperature and moisture characteristics. There are four basic types of air mass: tropical maritime, tropical continental, polar maritime and polar continental. Additional extremes are Arctic maritime and Antarctic continental. Continental air masses are relatively dry and maritime air masses are relatively humid.

Polar maritime air is common in both hemispheres with source regions in the high-latitude oceans. Polar continental air occurs in the northern hemisphere while Antarctic continental air occurs in the southern hemisphere. Tropical maritime air is common in both hemispheres, but tropical continental air is less common due to the lack of large land masses in the subtropics (e.g. North Africa is the main one). India can be a source region for tropical continental air in winter. In the summer the high pressure over Siberia combined with the Himalayan mountain chain restricts the movement of tropical continental air northwards.

Air masses are modified by the Earth's surface. If the surface is colder than the air mass then the stability of the air will be increased. If the surface is warmer than the air mass then stability will be decreased and cloud formation and precipitation may result. For example, when Arctic maritime air moves south over the North Atlantic the sea surface is warmer than the air mass. This warms the lowest layers, decreasing stability, encouraging convection and the formation of frequent showers. Weather forecasters study the source areas for air, track their movement over a particular region and predict how the air masses might be modified by local conditions and how they might interact with other air masses.

The climate zone of the mid-latitudes is often split into two regions: the first is the western edges of the continents; the second is the interiors of the continents combined with similar conditions experienced on eastern edges of continents. The winds experienced on the eastern continental edges of the mid-latitudes (e.g. eastern side of North America and Asia) have generally had a long journey across land. Hence the climate is closely related to that in the centre of the continent. Therefore, these two areas are classified under the same climate zone.

Mid-latitude western continental edge climates are mainly found in the northern hemisphere but New Zealand, Tasmania and southern Chile are also considered to fit this zone. The western margins of continents have mild winters for their latitude because of the warm ocean currents (e.g. Gulf Stream or East Australian Current as shown in Figure 1.3). These climates have a small range of annual temperature with precipitation distributed throughout the year but with considerable enhancement of precipitation by coastal mountain ranges (see below). Average winter temperatures are typically between 2° and 8°C with average summer maximum temperatures between 15° and 25°C. Precipitation totals generally range from 500 to 1,200 millimetres per year. These climate zones can be windy, particularly in coastal areas, and mid-latitude depressions can bring strong, damaging winds. A Mediterranean-type climate, consisting of a mild, half-year wet winter and a half-year hot, dry summer is found in south-west South Africa, central Chile, on south-west coastlines in southern Australia, in California, as well as in the Mediterranean itself (see also Chapter 5 for a description of the typical vegetation of Mediterranean-type climate areas). Average winter temperatures in Mediterranean climates range from 5° to 12°C with summer daytime maximums of 25° to 30°C. Precipitation totals typically range from 400 to 750 millimetres per year but with a summer minimum.

The mid-latitude continental interiors and their eastern margins have cold winters (average winter temperatures around 0°C) with frequent snowfall which often does not melt until the spring thaw while summers are hot and humid (around 25°C). Winters become colder further north (of 45°N) or west into the centres of the mid-latitude continents and summers also become cooler and less humid. North of 50° there are severe winters and relatively short summers (maximum of three months with average temperatures above 10°C). Precipitation tends to be distributed throughout the year in the mid-latitude continental interiors but generally with a summer maximum (mainly from convective showers and weak frontal systems). Total annual precipitation is low (below 500 milli-metres), but the cold winter and summer precipitation peak pro-vides sufficient moisture for plants (e.g. the main wheat-growing areas of North America). The regions above 50° in the mid-latitude continental interiors are dominated by high pressure in winter but mid-latitude weather systems at other times of the year. Winter average temperatures can be less than −25°C in the coldest month and are as low as −50°C in Siberia.

Tropical and subtropical climate and weather

Close to the equator the Coriolis effect is negligible and so the weather is not dominated by the movement of large circulatory weather systems. Here, air simply flows from high to low pressure. However, flow from the north-east and south-east trade winds helps air to converge into the intertropical convergence zone (see above). Air meeting at this zone and the warm conditions forces upward movement and the formation of clouds. Close to the equator the weather is dominated by frequent convectional clouds and plentiful rain with some parts of Amazonia and West Africa receiving over 4,000 millimetres per year. Average temperatures in the equatorial regions are around 27°C throughout the year. Equa-torial climates are not the same in each place, however, as topogra-phy and proximity to the oceans play a role in altering temperature and rainfall.

Moving north or south away from the equator into the trade wind region, between about 5° and 20° latitude, a rainy (summer) and dry (winter) season becomes clearer. The trade winds produce a steady, but not particularly severe, wind regime. However, tropi-cal depressions can form over the oceans, some of which become tropical cyclones. These cyclones are known as hurricanes in the Atlantic and typhoons in the west Pacific. Tropical cyclones require high sea surface temperatures (at least 27°C) but do not occur close to the equator as the Coriolis force is too weak. They only form over oceans because the energy to maintain their strength comes from the heat released when water vapour condenses to form clouds. When the moisture source is restricted as the cyclone hits land then the storm dies down. Tropical cyclones have high rainfall intensities and, as they travel slowly, any one place by which they pass can receive very high rainfall totals. The rainfall over two to three days from a single hurricane can be several hundred millimetres.

The climate of the trade wind belt also includes the monsoon which occurs in Asia, west and east Africa, Australia and a weaker version over south-western USA. The monsoons produce regions that experience an exceptionally wet rainy season. Monsoon regions are all associated with the switching of the wind direction. In winter, winds blow off the relatively cold continent toward the warm ocean (warmer air rising above the oceans draws in air to replace it from the nearby continent). There are therefore stable, dry conditions over land. However, as the land warms in the summer then the wind reverses due to the low pressure (rising air) at the surface that forms over the warmed continent. Air from over the ocean flows toward the low pressure over land bringing with it lots of moisture. These changes are also supplemented by changes in the location of the jet stream above. In Asia the dry northerly wind over India reverses direction in May/June, and warm, humid air from the Indian Ocean flows from the south until around October, bringing torrential rains. The rains are not continuous but in some mountainous places the rainfall can be 10,000 millimetres per year. Monsoons are not always the same each year and El Niño years in Asia can be associated with the failure of the monsoon rains sometimes leading to crop failure.

In contrast to the wet conditions near the equator, the major deserts are commonly found around 30° latitude coinciding with the zone of descending, dry air from the Hadley circulation cell. The driest hot deserts are found in the western coastal regions of the continents where the subtropical anticyclones are most intense. The main features of hot desert weather are a wind which increases aridity and high daytime temperatures (often over 35°C). The dry air and clear skies produce large daily ranges of temperature (as much as 20°C in some places) and night temperatures can even drop below freezing in places.

Mountain climate and weather

Hills and mountains can substantially modify regional and local climate. There tends to be a greater amount of the Sun's energy received in the form of ultraviolet energy in mountainous areas. This can be harmful to humans, causing skin cancer if frequent, high doses are received. Pressure and temperature normally fall with increasing altitude within the troposphere. This means that typically it becomes cooler as you ascend a mountain. The higher and more isolated a mountain the more the air temperature will resemble that of the free atmosphere rather than the atmosphere heated a short distance from the Earth's surface.

If there are light winds and clear skies at night, this can allow the ground to cool forming a strong temperature inversion, drawing air down the slopes into cooler valley bottoms. If there is cloud cover or strong winds then this process will be restricted. During the day if the Sun's energy is strong and slopes are steep sided, the air can be warmed not only from below but from the sides of the steep slopes too. This leads to local winds and rising air up the slopes of some valleys. It is these features that are often util- ised by thrill-seekers who go hang-gliding or parasailing in hilly areas.

Precipitation totals tend to increase with altitude. However, this increase is greatest in mid-latitudes. In the tropics the increase in precipitation with height is more complex and often stops at around 1,500 metres. As air moving over the land surface reaches the mountain range it is forced to rise over the mountains. The cooling of the air that occurs as it rises reduces the temperature to the dew point. Further rising air leads to the formation of cloud and precipitation. The reasons for differences in response in mid- and low latitudes are not yet clear and research is hampered by the lack of monitoring data at high altitudes.

If the rising air over mountains has been forced to the dew point and cloud formation and precipitation has occurred, it means that when the air falls again, once it passes the mountain range and warms, then it will be much drier than it was before it encountered the mountain range. Therefore, in the lee (downwind) of moun- tains the climate can be significantly drier than on the windward side of the mountains. Since mountains are regular barriers to air flow they are also regular locations for heavy precipitation. In fact on the island of Kauai (part of the Hawaii chain of islands) in the Pacific, Mount Waialeale receives an amazingly high average of around 12,500 millimetres of rainfall per year on its wind-facing side. However, on the leeward side of the mountain, the slopes only receive 500 millimetres of rainfall per year. Nearby locations over the ocean where there are no mountains would normally receive around 640 millimetres of rainfall per year. The warm and dry wind that blows down lee slopes of hill and mountain ranges is called a **föhn wind**. These winds are increased further by the wave effect of fast moving air being forced over a mountain range, a little like flowing water over a pebble. There are different local names for föhn winds depending on where you are in the world, such as the Chinook in North America or Zonda in Argentina. These winds can be very important because when they start to flow they can increase local air temperature by up to 25°C in an hour causing sudden snow melt or avalanche risk and influencing plant growth. In fact in Canada, in the lee of the Rockies, temperature rises of over 20°C have been recorded in just a few minutes as the wind begins to flow. This can bring plants out of their winter dominancy only to be damaged when cold weather returns once the föhn wind has stopped. As the winds are warm and dry they may even increase fire risk.

Normally surface friction reduces wind speeds by about 30 per cent. However, exposed peaks and ridges can have higher wind speeds as there is less surface friction around them and so the wind experienced in mountains can be different from that in the low-lands because of the nature of the terrain (and not necessarily because of the altitude). Furthermore, if wind is funnelled through gaps or valleys between individual peaks then it can be more intense at those locations. The winds in mid-latitude mountains are also influenced by the prevailing westerly winds and these winds are generally faster at higher altitudes in the troposphere. However, in the tropical and subtropical trade wind belts, the north-east and south-east trade winds generally weaken with height. Therefore, wind speeds can be low on tropical and subtropical mountains.

More precipitation may fall as snow in mountain regions which can accumulate over time. Box 1.3 describes rain and snow forma-tion. Often snow melt in the spring can produce large seasonal river flows. Where there are glaciers these can modify the local climate. Glaciers cool air in contact with the surface and depending on the moisture content of the air they can act as either a local moisture source or sink. For example, if the air is warm then water vapour can undergo **sublimation** (change directly from the solid to gaseous state) into the air above the glacier and this uses energy and thereby cools the air.

Box 1.3 Rain and snow

Clouds contain tiny droplets of condensed water. The droplets are very light and remain suspended in the air. As these droplets move around some of them collide and join together. As they grow bigger they become heavier and can start to fall as the gravitational pull drawing them down will be greater than the force of the rising air keeping them buoyant. This process is typical in warm clouds and the deeper the cloud the bigger the drops will grow and the faster the rainfall.

In the middle and high latitudes, however, many clouds form where it is well below 0°C. Here, clouds contain ice crystals as well as water droplets. The latter exist because the very small size of the droplets means they do not freeze straight away and they become 'supercooled'. A peculiar property related to the lower saturation point of ice compared to that of water means that some of the water droplets evaporate and then freeze onto the ice crystals causing them to grow. As they become heavier and fall they also collide and stick to each other, forming snow. As the flakes fall they may warm and melt and then produce rain. Clearly, if the air is cold near the surface then this melting does not occur and snow will reach the ground.

Land and sea breezes

Water is slow to heat up and cool down. Therefore, deep water bodies witness little daily change in surface temperature. The air over these water bodies does not experience large daily fluctuations in temperature. Over the land there are greater daily changes in air temperature, particularly in the summer when the Sun's energy is stronger and the land heats up during day and cools down at night. In the middle latitude summer half-years, sea surface temperatures (and the air in the layers close to the surface) are cooler than land surfaces during the day and they are warmer than land surfaces at night. The same is true in the high latitudes but in the winter half-year the temperature of the snow-covered land may remain colder than sea surface temperatures during both day and night. Differences in local temperature across land and water result in sea and land breezes. These breezes can also develop where there are large inland bodies of water such as the Great Lakes in North America or Lake Victoria in Africa.

Sea breezes only form when there are light wind conditions during typically anticyclonic conditions. The sea breeze can exist from the surface to 2 kilometres above ground and even up to 100 kilometres inland and brings cooler, more humid air inland. These breezes can bring welcome relief in the tropics providing a fresher feel to the climate of the coastal areas. When the more humid air begins to warm over land it can start to rise forming clouds, which is why on some summer days it can be a cloudless day inland, but when you reach the coast it can be disappointingly cloudy. Sea breezes die off as night falls and are sometimes replaced by a weak land breeze blowing out to sea.

Vegetation and climate

If the land is bare then the air temperature can be affected by the darkness of the soil and the moisture content. Darker soils absorb more of the Sun's energy and therefore radiate more long-wave energy back into the atmosphere to warm it from below. Moist soils will facilitate more evaporation thereby cooling the atmosphere as energy is used to evaporate the water.

The effects of vegetated surfaces on local climate are complex. There can be large differences from forest to grassland and from field to field depending on the crop. Different plants reflect more or less of the Sun's energy into space. Wind speeds can be affected by the height and density of vegetation. In fact, humans often plant hedgerows or rows of trees where they require some shelter from the wind. Often temperatures are warmest during the day near the top layer of vegetation if that catches most of the Sun's energy. The vegetation acts rather like the ground surface and absorbs short-wave energy and releases long-wave energy to heat the atmosphere. Also, the very top of the plant canopy will not be the warmest place because wind will assist the release of water from plants as water vapour and result in cooling. The temperature and wind effects are amplified in forests and it can be much cooler in forests during the day than outside of the forest.

Urban climate

Urban areas often have a different climate from that found immediately outside the urban area. This is because of the nature of the building and road surfaces. It was noted above when we discussed mountain environments how surface friction can slow down wind. Urban areas tend to have much more 'rough' and variable terrain than surrounding rural areas and this creates friction and reduces wind speed, which can also mean that air pollution (e.g. from cars) within urban areas is not dispersed very quickly. However, in strong winds tall buildings create powerful gusts as the wind is squeezed through the gaps between them, sometimes making it difficult to walk or open doors.

Urban building materials tend to be good absorbers of the Sun's energy and do not reflect as much as non-urban land surfaces. Furthermore, because humans heat the local environment (within houses, offices, shops, industrial premises etc.) then energy can be released into the urban environment, especially in the winter in mid- and high latitudes. Air pollution in urban areas can also add to increased temperatures by increasing the concentration of greenhouse gases. There also tends to be less vegetation cover or open soils within urban areas and so evaporation is reduced. Evaporation normally causes cooling since energy is used up to transform water into water vapour. Hence if evaporation is reduced there will be less of this cooling effect compared to surrounding rural areas. The above factors combine to create the 'urban heat island' effect where the urban area acts as an island of warmer air surrounded by cooler rural air. The heat island is more apparent at night or when wind speeds are low. The heat island effect is related to the city size and types of building in the city. The largest heat islands are found in cities such as Beijing and New York and a heat island effect as large as 12°C has been found in Montreal during winter.

Because of atmospheric pollution in urban areas there can be periods when the weather conditions within urban areas are a danger to human health. In Britain the burning of coal for heat in urban areas during periods of calm weather in winter used to be associated with 'smogs'. Smoke particles mixed with fog give smogs a yellow colour. Smogs often settled over cities for many days. During smog periods people suffered respiratory problems and there were increased fatalities. One of the worst smog episodes in Britain occurred in London on 4 December 1952. The smog lasted for five days and resulted in approximately 4,000 more deaths than usual. In response, legislation was developed to create smokeless zones where coal burning was banned and this resulted in rapidly improved urban conditions. However, the increased number of vehicles within urban areas is now a widespread problem for urban air quality, especially in regions that receive high amounts of the Sun's energy such as Athens, Mexico City and Beijing. Here, the ultraviolet radiation from the Sun reacts with vehicle emissions and produces a photochemical smog which irritates the eyes, nose and throat and creates a hazy atmosphere. During the Beijing Olympic Games in 2008, officials were so worried about this potential effect that they banned vehicles from large parts of the city, moved whole factories, halted building construction and replaced coal boilers with gas ones. This improved air quality substantially and has had long term beneficial effects upon the city.

SUMMARY

- The climate system is a closely integrated system involving the atmosphere, oceans, land surface and vegetation.
- The Sun heats up the Earth's surface which in turn warms the atmosphere from below.
- Greenhouse gases trap some of this heat so that at night-time it does not all escape back into space.
- The energy surplus from the Sun at the equator and the deficit at the poles creates an imbalance that leads to heat transfers through air and water.
- The Earth's rotation has a large impact on air and water movements while the Earth's tilt causes seasonal differences in the receipt and redistribution of the Sun's energy.
- There is a zone of moist, warm ascending air near the equator which yields a plentiful rainfall supply.

- Tropical cyclones can form over warm oceans and bring large amounts of rainfall and damaging winds to coastal areas.
- At around 30° north and south there are major zones of descending air which are clear and dry and this is where most of the world's large deserts are found.
- At the surface in middle latitudes, the main features are easterly moving cyclonic and anticyclonic systems which bring highly variable weather conditions.
- Continental interiors have different climates from those areas close to the oceans and ocean circulation which transports heat energy can have a major influence on the climate of a region.
- Polar climates tend to be cold and dry, being associated with descending air and clear conditions, but they can be very windy.
- Topography plays an important role in modifying the local and regional climate; upland areas are often more windy and receive more precipitation. However, the downwind side of mountain areas can be very dry.
- Sea and lake breezes can influence the local climate of areas near water bodies.
- Urban environments tend to be warmer than rural environments due to building materials that absorb the Sun's energy, air pollution that traps more heat, local heat escaping from buildings which humans have heated and reduced evaporation.

FURTHER READING

Five books that provide excellent in-depth coverage of material covered in this chapter with plentiful illustrations are:

Barry, R.G. and Chorley, R.J. (2009) *Atmosphere, Weather and Climate*, ninth edition, London: Routledge.

Bridgman, H.A. and Oliver, J.E. (2006) *The Global Climate System: Patterns, Processes, and Teleconnections*, Cambridge: Cambridge University Press.

Hughes, K.K. and Mayes, J. (2004) *Understanding Weather: A Visual Approach*, London: Hodder Arnold.

Pinet, P.R. (2009) *Invitation to Oceanography,* fifth edition, Sudbury, MA: Jones and Bartlett Publishers Inc.

Sverdrup, K.A. and Armbrust, E.V. (2009) *An Introduction to the World's Oceans*, tenth edition, London: McGraw-Hill.

2

CLIMATE CHANGE AND CARBON

CLIMATE CHANGE

Long-term climate change

When you examine the scientific evidence it becomes clear that the Earth's climate has changed throughout its 4.6 billion year history, but usually on long, slow timescales. It may surprise you to learn that today we live in an ice age. This is evident by the fact that there are still large glaciers and ice sheets covering parts of the planet. These ice sheets have been present over the past 2.4 million years and have expanded and contracted in cycles. This period of time is known as the **Quaternary**. However, before the Quaternary period the Earth was much warmer and lacked ice sheets. In fact, it was warm for around 280 million years. Further back in time there were four other ice age periods in Earth's history which lasted for a few million years separated by very long warm periods lasting hundreds of millions of years. So while our current Quaternary ice age sounds like it has lasted a long time it is actually very short compared to the timescales of Earth's history. Understanding past changes to the Earth's climate provides information which is useful to our understanding of our current climate and what might happen in both the near and long-term future.

The start of the Quaternary period seems to have been initiated by **plate tectonics** (see Chapter 3). The movement of the continents created the suitable conditions for the beginning of the ice ages by positioning Antarctica over the South Pole, thereby allowing a large land mass to cool and build up an ice sheet which then further cooled the climate by becoming a large reflective body of the Sun's energy. The northern hemisphere continents were also huddled around the Arctic Ocean and the ocean circulation systems were fixed into new positions. This then allowed other climate forcing factors to become important.

During the Quaternary the climate has cooled and warmed many times and this has been associated with major advances and retreats of ice sheets. This has shaped the land surface, carving new features out of rock, depositing sediments and landforms and therefore leaving evidence of past climates behind. When ice sheets grow sea levels fall because water is taken out of the oceans and locked up on land. The last major ice advance peaked at around 18,000 years ago and then retreated. At this time it was possible to walk from Britain to Europe across land because the North Sea and English Channel were dry since sea levels had dropped by 120 metres. The last 10,000 years have been relatively warm and this warm period within the Quaternary is known as the **Holocene**. The cold periods when ice advances are known as **glacials** and the warmer periods in between are known as **interglacials**. The Holocene is our present interglacial. Several processes have been proposed to explain Quaternary climate changes and many of these involve internal feedback mechanisms within the Earth's climate system. However, it has been shown that changes in the receipt of the Sun's energy on Earth helps to partly explain the glacial cycles that have occurred during the Quaternary. This is known as orbital forcing or the 'Milankovitch theory' which is named after an early twentieth century mathematician who examined the Earth's orbit around the Sun.

Orbital forcing theory is based on the idea that the amount of energy reaching different parts of the Earth from the Sun varies through time, but in a regular and predictable way. It varies with three factors as shown in Figure 2.1. First the shape of Earth's orbit around the Sun (its eccentricity) varies over around 100,000 year cycles as it moves from being more circular to more elliptical and back again. Second, the Earth's axis around which it rotates is currently tilted at 23.5°. But this tilt varies through time from 21.8° to 24.4° over a 41,000 year cycle. Third, the Earth has a slow wobble on its spinning axis caused by the gravitational pull exerted by the Sun and the Moon. This happens over two cycles of 19,000 and

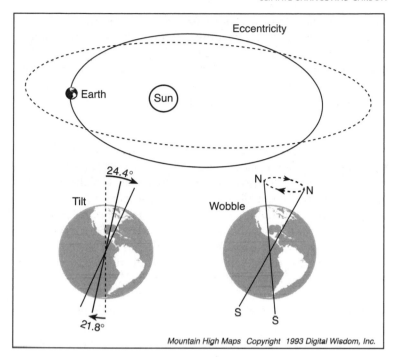

Figure 2.1 The three slowly changing and cyclical mechanisms related to the Earth's orbit around the Sun that impact the energy that reaches the Earth's surface.

23,000 years. The eccentricity effect causes the seasons in one hemisphere to become more intense while the seasons in the other are moderated. The greater the tilt effect the more intense the seasons in both hemispheres become; summers get hotter and winters colder. The wobble determines where in the orbit the seasons occur, and most importantly the season when the Earth is closest to the Sun. The Earth currently reaches its furthest point from the Sun during the southern hemisphere winter. Therefore, southern hemisphere winters are slightly colder than northern hemisphere winters while southern hemisphere summers are slightly warmer.

It is remarkable that scientific evidence from marine sediments and other sources shows that the timescale and frequency of the advance and retreat of ice sheets produces a close match to cycles of solar energy predicted by orbital forcing theory. For example, there have been eight large glacial build-ups over the past 800,000 years on an approximately 100,000-year cycle each coinciding with minimum eccentricity. Smaller decreases or surges in ice volume have come at intervals of approximately 23,000 years and 41,000 years in keeping with the frequency of the other two orbital mechanisms.

While orbital forcing theory matches the timing of the cool and warm periods it is not sufficient on its own to explain the magnitude of changes in temperature experienced. In fact there appears to be a 4° to 6°C shortfall. Furthermore, the orbital cycles mathematically predict a smooth rise and fall of temperature. However, the evidence for the actual build-up and retreat of ice shows a sawtooth pattern. Over tens of thousands of years ice sheets built-up several kilometres thick, souring and scarring the landscape as far south as central Europe and Midwestern USA (Figure 2.2). But each cycle ended abruptly. Within a few thousand years the ice sheets melted back to present-day patterns. So orbital forcing seems to be a driver of climate change through the Quaternary, but other feedbacks must be operating to deal with the temperature shortfall and the sawtooth pattern. Understanding these feedbacks might be important for us so that we may understand how the climate may evolve in the future.

One positive feedback comes from increased reflection (**albedo**) of the Sun's energy from the ice causing a further drop in temperatures, allowing ice sheets to expand further. This would be enhanced by sea level fall following ice sheet growth allowing ice sheets to expand further on land enhancing albedo. Nevertheless, these feedbacks are still not sufficient to explain the magnitude of temperature changes observed.

An additional feedback which has been the centre of research for the last two decades is related to changes in the thermohaline deep water circulation system in the oceans (described in Chapter 1) and the role of the oceans in changing atmospheric composition. Ice cores from the middle of the oldest ice sheets in Greenland and Antarctica contain bubbles of gas. These bubbles contain air from the time when the snow fell that later formed the ice. Data from bubbles contained within ice cores indicates that carbon dioxide

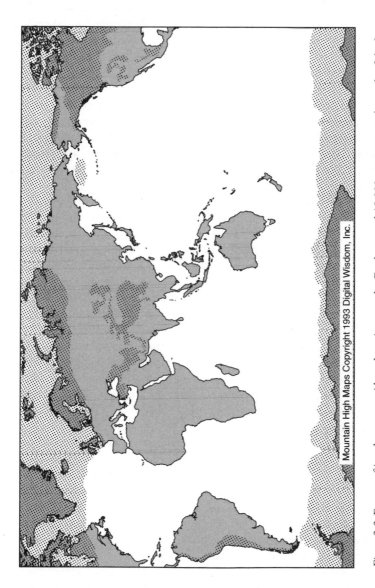

Figure 2.2 Extent of ice sheet cover (dotted area) across the Earth around 18,000 years ago at the peak of the last glacial cold period.

concentrations in the atmosphere were greater during interglacials and lower during glacials. As a greenhouse gas the changes in carbon dioxide concentrations in the atmosphere will have significantly affected the Earth's temperature. Carbon dioxide is exchanged fairly easily between the oceans and the atmosphere. If there are changes to carbon dioxide concentrations in the ocean surface waters this may affect atmospheric concentrations which may then cause the Earth's climate to warm or cool. Plankton (tiny green plants) in the upper ocean take up carbon dioxide from the water as part of photosynthesis and this is converted into the plant material composed of carbohydrates. Dead plankton and other marine waste may sink to the ocean floor moving the carbon into the deep. The thermohaline circulation system is driven by temperature and salt concentration gradients and acts as a strong, deep current that pumps carbon dioxide and nutrients from the surface of the oceans to the deeper waters and returns them to the surface again. There are sensitive zones where such downwelling and upwelling occurs (see Chapter 1). If the thermohaline circulation acts in its current fashion then the carbon dioxide on the ocean floors would be stirred up and taken back to the surface. Deep carbon stores will not be returned to the surface as quickly if the thermohaline circulation was to slow down. The surface of the oceans would therefore be depleted of carbon dioxide and less will be returned to the atmosphere. Overall this process would result in decline of atmospheric carbon dioxide concentrations as the ocean plants continue photosynthesis. The energy transfer rates between the equator and poles would also be affected by changes in the ocean circulation system.

There is evidence that the thermohaline circulation system slows down during glacials. One reason suggested for this is the change in local air currents around thick ice sheets (of several kilometres). Reduced evaporation in sensitive areas of deep water formation in the North Atlantic (i.e. wind enhances evaporation, which makes the water more saline which then makes it more dense and so it sinks) could slow down the rate of sinking or switch it off and so the whole deep ocean circulation system becomes more sluggish. Return of carbon dioxide from the deep to the atmosphere would slow, since the upwelling that occurs elsewhere would be reduced. The above theory is not entirely accepted but it does elucidate the current line of thinking and it is widely believed that the North Atlantic is a very sensitive part of the Earth's climate system and changes there can have global impacts. Once increased solar energy starts to warm the planet as we move out of a glacial period then positive feedbacks could result in the rapid warming indicated by the sawtooth pattern in the evidence. This may be related to lower albedo as ice retreats and the sudden switching on of the thermo-haline circulation system.

As well as the long-term cycles during the Quaternary there have also been shorter cycles some of which are just a few decades long. Many of these fast-changing periods, however, seemed to have occurred during glacials rather than interglacials. This was probably related to interactions between ice sheet dynamics, ocean circulation and biological productivity. So far we have only found evidence for a few rapid changes in climate during interglacial periods and it should be noted that we currently live in an interglacial. Therefore, the rapid climate changes that seem to be occurring at the present time are highly unusual. Any evidence for rapid climate change during interglacials of the past (e.g. around 8,200 years ago) is therefore being actively explored at the moment as these changes may provide a guide to possible climate behaviour in the near future.

Recent climate change

Over the last century glaciers and ice caps have been receding, snow cover has reduced and sea levels have been rising. The Earth has warmed by about 0.7°C, particularly over the last 15 years (Figure 2.3). The concentrations of greenhouse gases such as carbon dioxide and methane in the atmosphere are greater than at any time in roughly the last half million years and their concentrations have risen sharply in recent decades. The current concentration of atmospheric greenhouse gases is equivalent to carbon dioxide levels of around 430 parts per million compared with only 280 parts per million before the Industrial Revolution (this is calculated by determining what each greenhouse gas equates to in terms of its warming role and expressing it in units of carbon dioxide equivalents). The rate of increase in temperature is faster now than previously, even at the end of glacial periods. And we

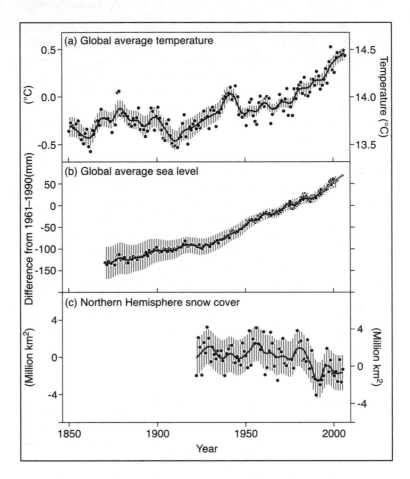

Figure 2.3 Observed changes in (a) global average surface temperature; (b) global average sea level from tide gauge (black line) and satellite (white dots) data; and (c) Northern Hemisphere snow cover for March–April. Smoothed lines represent decadal averaged values while black circles show yearly values. The vertical lines are the uncertainty intervals estimated from a comprehensive analysis of known uncertainties (a and b) and from the time series (c). Source: *Climate Change 2007: The Physical Science Basis.*

have to put this into the context of the fact that we are living an interglacial which is usually a settled period without sudden temperature changes. The Intergovernmental Panel on Climate Change (IPCC) stated in their report in 2007 (see further reading) that: 'warming of the climate system is unequivocal, as is now evident from observations of increases in global average air and ocean temperatures, widespread melting of snow and ice and rising global average sea level'.

The temperature increase is widespread over the globe but more exaggerated at higher northern latitudes. For the Arctic the temperatures over the last 100 years have increased at twice the rate of the global average. The oceans have warmed down to depths of at least 3,000 metres. Numerous other long-term changes in climate have been observed. For example, there have been increases in rainfall in northern Europe, northern and central Asia and eastern North and South America between 1900 and 2005. Rainfall has declined in the Sahel, the Mediterranean, southern Africa and parts of southern Asia. The IPCC noted in their 2007 report that it is very likely that cold days, cold nights and frosts have become less frequent over most land areas, while hot days and hot nights have become more frequent. It is also likely that both heatwaves and heavy precipitation events have become more frequent over most land areas.

Some have argued that these changes are naturally driven. For instance, solar cycles including sunspots, which are dark marks on the solar surface and occur in 11-year cycles, affect the energy we receive. However, the Sun's energy release between 1990 and 2010 was lower than normal and therefore goes against any observed rise in near-surface ocean temperatures. Volcanic eruptions eject greenhouse gases into the atmosphere but these have not become more frequent over the last 100 years and so cannot be the cause of the trend in global warming.

It is absolutely certain that humans have emitted large quantities of carbon dioxide by burning fossil fuels and vegetation. Similarly, the increases in methane concentrations (which are currently rising at faster rates than carbon dioxide) can be attributed to energy production, food production from rice fields or ruminant livestock, landfills, waste treatment and vegetation burning. Recent estimates suggest that one-third of nitrous oxide emissions are caused by human action; the rate of its production increases in agricultural areas when nitrogen fertilizer is used. Humans have also emitted fumes and smoke from industrial processes and transport. These **aerosols** partly shield the planet from the Sun's energy scattering and reflecting it. However, not all aerosols behave in the same way. Aerosols from vegetation burning, 'black carbon', and the 'brown clouds' that come from urban sources may have the opposite effect. The former seem to cause cooling while the latter result in warming.

Almost every time there is a flood or storm, the media seem to blame it on global warming. At the same time, the media love to create polarised storylines and so if there are some people, including scientists, who do not accept that climate change is being driven by human action the media will push their views into the spotlight. However, it needs to be made absolutely clear that the overwhelming scientific consensus is that the production of greenhouse gases by humans is the primary cause of recent global warming. Masses of evidence have been compiled by the IPCC which consists of climate experts from around the world who use all of the available evidence and have come to this conclusion. The IPCC have also shown that Global Climate Models, which simulate the production of greenhouse gases, show the same pattern of global warming as we have observed. When these models are run without adding human production of greenhouse gases, the models show no appreciable global warming (see www.ipcc.ch).

Predicting how the climate will behave in the near future is an important and growing research area. It is important to understand the feedbacks within the Earth's system in order to do this. Negative feedbacks reduce any warming effect while positive feedbacks cause accelerated warming. It is interesting to note that there are several possible major negative feedbacks which are being investigated for their role. For example, warming will lead to a more active hydrological cycle with more evaporation and rainfall. Rainfall contains small amounts of carbon dioxide which when it meets rock can erode it transforming the carbon dioxide into calcium carbonate which then gets washed into rivers and oceans. Enhanced rainfall and rock weathering will therefore have the net effect of slightly depleting the atmosphere of carbon dioxide. Increased evaporation and **transpiration** (water release from plants) in a warmer world will lead to more clouds, cooling the planet. Increased precipitation and the amount of fresh water release by melting ice could increase the volume of fresh water entering sensitive parts of the oceans thereby slowing the thermohaline circulation in the ocean and allowing northern high latitudes to cool. There are also additional negative feedbacks caused by human action on land. For example, deforestation may result in accelerated erosion of soils which could increase the dustiness of the atmosphere reflecting more of the Sun's energy back into space. Furthermore, if warmth-loving broadleaved forests replace coniferous ones, or if forests are replaced by agriculture, then the albedo will increase causing cooling.

Positive feedback processes are also currently being studied. For example, albedo will decrease due to the retreat of snow and ice cover. Warming may cause release of carbon dioxide from enhanced decomposition of vegetation, especially in forest regions and the tundra (but faster growth rates in these areas may also take up more carbon dioxide). Increased cover of woody vegetation in the high latitudes, caused by an increase in suitable growing land for plants due to warming, will decrease albedo. Warming may increase the decomposition rate of **gas hydrates** leading to a release of the potent greenhouse gas methane; this will increase warming. Gas hydrates are solid crystal structures in which the molecules of gas are combined with molecules of water. There is twice as much carbon stored in gas hydrates on Earth than on all other fossil fuels put together. The hydrate of methane is stable at low temperatures or high pressures (e.g. in frozen ground or the bottom of the oceans). However, global warming may melt some of the ground that is currently frozen in high latitudes potentially releasing the methane from the hydrates into the atmosphere. Some peatlands may warm increasing the rate of decomposition of the plant matter that forms them, resulting in more carbon being released than they store each year.

IMPACTS OF CLIMATE CHANGE

The above feedbacks could in themselves be seen as impacts of climate change. Most models predict a 2° to 4°C increase in global temperature over the next century and a 50 per cent chance that average global temperatures could rise by 5°C. Substantial impacts from such increases are anticipated. As Sir Nicholas Stern put it in his 2006 report for the UK government on the economic impacts of climate change, 'Such changes would transform the physical geography of the world. A radical change in the physical geography of the world must have powerful implications for the human geography – where people live, and how they live their lives'. Indeed a 2010 report for the US National Research Council (*Climate Stabilization Targets: Emissions, Concentrations, and Impacts over Decades to Millennia*) suggests that the Earth is now entering a new geological epoch called the Anthropocene during which changes to the planet's environment will be dominated by effects of human activities, notably emissions of carbon dioxide. Because carbon dioxide is long lived in the atmosphere then emissions now can lock the climate system into a range of impacts over many generations.

Key potential impacts include:

- There may be more severe weather events (storms, floods, hurricanes, heatwaves, droughts etc.) as a warmer Earth means there is more energy available for atmospheric and oceanic transfers. For example, there could be a 10 per cent increase in average hurricane wind speed.
- Warming may induce sudden shifts in regional weather patterns such as the monsoons or the El Niño Southern Oscillation. This could have major consequences for water availability, crops, flooding and wildfire in tropical regions. Some climate models predict a more frequent and intense El Niño, perhaps even causing replacement of the rainforest of Brazil by **savanna**.
- Melting glaciers will at first increase flood risk and then once depleted reduce water supplies. Large parts of the Indian subcontinent, China and the Andes rely on summer glacial meltwater to provide water for crops and to drink. Without this regular supply new ways will have to be found to capture water for around 1 billion people.
- 40 per cent of the world's population live within 100 kilometres of the coast and two-thirds of the world's largest cities are found there. Rising sea levels from ice melt and the expansion of the water upon warming could result in hundreds of millions of

more people enduring floods each year. More than a fifth of Bangladesh could be submerged by 2100.

- Some cold areas such as Canada, Russia and Scandinavia may enjoy an increase in agricultural productivity as a result of warming, at least in the short term, while many of the world's main food-producing regions may become too hot and dry for crops to grow (e.g. central and southern Europe and the Great Plains of North America). Declining crop yields, especially in Africa, could leave hundreds of millions without the ability to produce or purchase sufficient food.

- The area burnt by wildfires may increase substantially. It has been estimated that the area burnt by wildfire in western United States may increase by 200 to 400 per cent for every degree of warming.

- Deaths from malnutrition and heat stress will increase and diseases such as malaria could become more widespread. Diseases are likely to spread from the tropics to the mid-latitudes as the climate warms and outbreaks of pests may become more extreme because natural biological controls may be lacking in these newly colonised pest areas.

- Plant and animal extinction rates will be increased as many species cannot migrate fast enough to keep up with the temperature changes. Up to 37 per cent of species may be on the road to extinction by 2050. Further climate change impacts on ecosystems are discussed in Chapter 5.

- Ocean acidification is occurring and will increase as the oceans absorb carbon dioxide from the atmosphere. This buffers global warming but it can have major effects on marine ecosystems. For example, it reduces the ability of species such as molluscs, crustaceans, coccoliths and corals from forming the calcium carbonate needed for their structures and this would have knock-on effects for the food chain.

- Global economic output could be reduced by 3 to 10 per cent with the poorest countries most badly affected.

THE CARBON CYCLE

Carbon is an element present in air and rocks, is dissolved in fresh and ocean waters, and occurs in all living matter. It is the fourth most common element in the universe and is essential for life. Carbon moves through and across the Earth in several forms and via a number of processes in what is known as the carbon cycle (Figure 2.4). The carbon in your body has been recycled millions and billions of times forming parts of lots of other organisms, rocks, water bodies and the atmosphere. The natural carbon cycle involves the transfer of carbon from the atmosphere to the land and ocean where it remains in forms such as living matter, rock, soil or dissolved compounds, before eventually returning to the atmosphere. The total amount of carbon on Earth remains the same through time, but the same carbon exists in different forms. The carbon cycle operates over both long (millions of years) and short (hours/days) timescales. There is much variability in how long carbon will be stored in particular locations but on average a carbon atom will stay in the atmosphere for five years and in the oceans for 400 years.

One of the main mechanisms for the transfer of carbon out of the atmosphere is plant photosynthesis whereby the Sun's energy helps plants transform carbon dioxide into carbon and oxygen. The carbon helps form the structure of the plant material while the oxygen is immediately released into the atmosphere. Most biological matter is about 40 to 60 per cent carbon when it is dried and the constituent parts are weighed. Carbon is transferred when animals eat plants (see Chapter 5). Respiration uses oxygen to break down organic matter into carbon dioxide and water. Unlike photosynthesis (which only occurs in plants) respiration occurs in plants and animals. For animals, respiration is the breakdown of food by oxygen to release the energy stored by the food and this produces carbon dioxide. Death of plants and animals means that carbon can be returned to the ground. Some of this is very quickly returned to the atmosphere by fire or decayed by bacteria which form carbon dioxide, whereas some may be stored in the soil for longer periods. Temperature is an important factor controlling the carbon cycle. Rates of decay are much faster in tropical environments and slower in cold environments. The amount of carbon that is lost from the atmosphere through photosynthesis and then released back to the atmosphere is very large. This means that the annual carbon dioxide concentration in the atmosphere fluctuates during each year according to peak periods of growth and decay.

The fluctuation effect is dominated by the northern hemisphere forests. Since there is more land mass in the northern hemisphere and forests are concentrated here then there is much more carbon uptake in the northern hemisphere summer and therefore global atmospheric carbon dioxide concentrations decrease during each northern hemisphere summer and rise each winter.

As discussed earlier in this chapter, the oceans also absorb carbon dioxide from the atmosphere via photosynthesis by microscopic plants such as algae. This photosynthetic activity occurs mainly in the upper 50 metres of the ocean but varies widely across the oceans depending on temperature and the amount of nutrients available in the water to support life. While most of the carbon is returned to the atmosphere as part of the cycle of life some of it falls to the deep ocean floor within the detritus of dead plant and animal material. Part of this carbon can dissolve into the seawater whereas some of the carbon can remain stored for thousands or even millions of years on the ocean floor. However, deep ocean currents can stir up the sediment and bring some of the carbon back to the surface to later be released back to the atmosphere. The oceans currently store around 60 times more carbon than is in the atmosphere and 20 times more than in the ecosystems and soils on land.

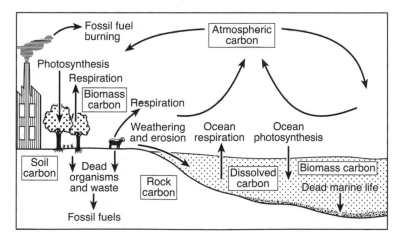

Figure 2.4 Major components of the carbon cycle.

Over timescales of millions of years, rock weathering (as described above and see also Chapter 3) adds carbon to river water flowing into the oceans. This carbon can be extracted by marine animals for their shells and bones and when they die the carbon in the sediment can drift to the ocean floor. Some of this can be returned to the ocean surface by deep upwelling currents, but some can build up as a deep deposit. The slow movement of the Earth's plates means that ocean floors eventually get swept under the continents (see Chapter 3 for an explanation) and when this happens the sediment can be heated, melted and released back into the atmosphere (e.g. during volcanic eruptions). Additionally, carbon can get locked away within the Earth's surface for a long time when animals and plants die and the carbon enters the ground (e.g. in a peatland). Over long time periods this deposit can get buried and compressed and form coal or oil. Human extraction of this 'fossil carbon' for fuel then releases the carbon back to the atmosphere when it is burnt.

Currently our burning of fossil fuels releases approximately 6.5 gigatonnes (thousand million tonnes) of carbon per year. Deforestation in the tropics releases a further 1.5 gigatonnes per year. However, the sums do not quite add up since atmospheric carbon dioxide concentrations are only increasing by 3 gigatonnes per year. It seems that the additional carbon is being taken back up by the oceans or by additional photosynthesis with the latest data suggesting about half of this additional carbon for each. However, these additional sinks for carbon might not continue to absorb such a high proportion of the carbon we have been releasing into the atmosphere for much longer. There is probably a threshold maximum amount of additional carbon that the oceans and vegetation can take up and once this buffer has been used up the rate of change in temperature might climb even further. As an alternative to fossil fuels the use of biofuels has been proposed whereby plants are grown to be used as fuel. The process is meant to be 'carbon neutral' in that the same amount of carbon released into the atmosphere from burning was taken up by the plant in its growth. However, as Box 2.1 illustrates, the story is not simple.

Box 2.1 Palm oil biofuel and climate change

Biofuels are seen by many to be good for reducing greenhouse gas emissions and reducing the magnitude of climate change. This is because they involve burning plant materials which are then replaced with new plants or by new fruits and so over a period of a few years the same amount of carbon is taken out of the atmosphere as is released back again. One popular biofuel which is also used in food products is palm oil. Palm oil is produced from the pulp of the fruit of the oil palm plant. It is used to make forms of biodiesel which is becoming increasingly popular in Europe. Because of its popularity for export, the governments of Malaysia and Indonesia are encouraging the planting of more and more oil palm trees and production of palm oil.

Much of this new planting is taking place on land which is covered by thick deposits of tropical peatlands. These peatlands are large carbon stores which have been amassed over thousands of years. However, to plant oil palm trees on peatlands the peat needs to be drained. Around 45 per cent of the 27 million hectares of peatland in these two countries has now been drained. The drainage removes waterlogging, the lack of which encourages faster decomposition of the peat and release of the carbon that it had been storing. Such carbon emissions from drained peatland has played a major contributing role in moving Indonesia to being the third largest greenhouse gas producer (through human action) after the USA and China. Thus a strategy that is supposed to be supporting low carbon fuel production is actually releasing large quantities of carbon into the atmosphere.

CLIMATE CHANGE MITIGATION AND ADAPTATION

Sir Nicholas Stern reported in 2006 in his review of the economics of climate change that '[t]he scientific evidence is now overwhelming: climate change presents very serious global risks, and it demands an urgent global response'. The overall argument presented in the report was that action now would cost a lot, but this would be a lot less costly than if we did not take action given the nature of the impacts described above. So what can be done to combat climate change?

Greenhouse gas emissions can be reduced by improving the efficiency of global energy supply, reducing demand for goods and services that emit a lot of greenhouse gases in their production, avoiding deforestation, peat drainage and other land management activities that increase emissions and using lower-carbon techniques for power or heat generation and transport. Of course, the above can only be fully realised if the international community works together. If only one country acted then it would be at an economic disadvantage while the climate system would have minimal benefit. There have been attempts at getting the international community to sign up to legally binding agreements and the 1997 Kyoto Protocol was ratified by 175 countries (but not by major emitters such as USA, India and China). The commitments added up to a total cut in greenhouse gas emissions of 5.2 per cent by 2012 from 1990 levels. Only a few countries have met their targets and unfortunately a 5.2 per cent reduction is inadequate to prevent 'strong' global warming. The December 2009 Copenhagen summit sought to get agreements from more countries for tougher reductions but these talks failed to reach an agreement.

If countries are to reduce their greenhouse gas emissions then economists such as Nicholas Stern have argued that there should be an appropriate common global price put on carbon. This is why you may have heard the term 'carbon trading'. Those who are polluting are causing consequences for others who will have to pay the cost. Therefore, it is argued that the polluters should pay this cost. The purpose of this would be to encourage people to move to low carbon technology and be more efficient about the carbon they use. It may also help pay for investment in further mitigation strategies. Such strategies might include providing better insulation for homes so that less energy is wasted, or moving to more wind, wave or nuclear energy generation methods. Other strategies include capturing carbon from power stations and other industrial plants and pumping it into natural underground gas stores or disused oil wells. This might sound fanciful but it is really taking place. There are hundreds of planned sites across the world and several operation sites for carbon capture and storage such as at locations in the North Sea, at In Salah in the Algerian central Sahara or the Lacq gas processing plant, south-west France. Scientifically, there are also prospects of enhancing the strength of the global carbon sinks in order to slow down the rate of climate warming. Planting more forest or protecting existing tropical forests, peatlands and other wetlands will be important. However, as the world's population grows then demand for land for agriculture will rise making such protection challenging. The ocean carbon sink might also be managed. This could be done by fertilising the ocean to provide more food for plankton which can then expand their populations. This would lead to more photosynthesis and also create a more vibrant food chain with increased deposition of dead biota, which is rich in carbon, to the deep ocean. However, this is ethically challenging.

As well as mitigation, adaptation is needed to deal with the inevitable impacts of climate change that will occur before mitigation measures take effect. For example, adaptation measures include moving people away from flood plains that are more likely to be flooded in the future or designing better disaster management and emergency response procedures. Investment will be needed in the poorest countries.

Individually, some people try to take personal responsibility for their carbon emissions by calculating their carbon footprint and changing their lifestyles or paying for carbon offsetting schemes. The **carbon footprint** is the amount of carbon (often expressed in terms of the equivalent amount of carbon dioxide) that an individual (or organisation) uses over a given time period. It is fairly easy to work out the carbon footprint of energy use or travelling by car or train as emissions can be calculated per person. However, what if you purchase some clothing or food that has been produced by industrial methods and transported long distances? It is often more difficult to work out how much 'embedded carbon' was used in making or transporting that product. Furthermore, it is not clear how you would know whether or not you have already paid for that carbon through the price you paid for the product (i.e. has the company already paid a carbon tax for the carbon it used in making the product and therefore passed that cost on to customers). This is why more clarity is needed on global carbon pricing so that organisations and individuals have more long-term confidence and understanding of the carbon payment system. Carbon offsetting involves paying an organisation for carbon savings they have made. For example, there are organisations who plant trees to store carbon from the atmosphere and then seek voluntary payments from people who want to match their carbon footprint with a carbon saving made as part of the tree planting scheme.

SUMMARY

- The Earth's climate has changed significantly through its history.
- Climate changes over the past 2.4 million years have been dominated by cycles of cold and warm during which major ice sheets expanded and contracted again.
- The driver of historic climate changes over the Quaternary has been orbital forcing, although this is not sufficient in itself to explain the magnitude of changes that have been experienced.
- Internal feedback processes are important for reinforcing or dampening climate change; of particular importance are the ocean circulation systems.
- Climate change over the past century has been more rapid than at any time over the past half million years.
- Humans have emitted large quantities of greenhouse gases into the atmosphere over the last 200 years and continue to do so at a growing rate.
- It is expected that the Earth's climate will warm by between 2° and 4°C by 2100.
- The impacts of climate change on the physical geography of the Earth will be enormous.
- The impacts on human society will be equally large.
- More urgent and internationally co-ordinated climate change mitigation measures are needed if we are to avoid extremely serious climate change.
- Adaptation measures need investment and development now to deal with the impacts of climate change that will happen as a result of emissions we are already committed to.

FURTHER READING

Good texts on the climate changes of the past 2.4 million years include:
Anderson, D.E., Goudie, A.S. and Parker, A.G. (2007) *Global Environments Through the Quaternary: Exploring Environmental Change*, Oxford: Oxford University Press. A clearly written and well illustrated textbook.

Paillard, D. (2001) Glacial cycles: toward a new paradigm, *Reviews of Geophysics*, 39(3): 325–346. This review article explains extremely well the orbital forcing theory and the need to understand feedback mechanisms in the Earth's climate system.

Walker, M. and Bell, W. (2005) *Late Quaternary Environmental Change: Physical and Human Perspectives*, Harlow: Pearson Education. Clearly written book focusing on the last 20,000 years during which time ice sheets have retreated.

On contemporary climate change the following are useful:

Houghton, J. (2009) *Global Warming: The Complete Briefing*, fourth edition, Cambridge: Cambridge University Press. A very readable account of global warming which includes the latest IPCC figures and predictions.

Intergovernmental Panel on Climate Change (2007) *Climate Change 2007*. This is the IPCC report itself which comes in several volumes including summaries. These reports can be downloaded at www.ipcc.ch.

Stern, N. (2007) *The Economics of Climate Change*, Cambridge: Cambridge University Press. This is a thought provoking summary of the predictions for climate change over the coming decades and the potential impacts, demonstrating an economic case for urgent co-ordinated action.

3

TECTONICS, WEATHERING, EROSION AND SOILS

On a large scale the landscape and oceans change slowly through time, mainly through tectonic processes, weathering and erosion. These slow processes also bring sudden change and hazards associated with them. Movements in the Earth's plates result in sudden earthquakes or volcanic eruptions. Landslides are local hazards but are natural processes of change and therefore need to be understood within this context. Humans also modify the environment and can enhance hazardous risks of rapid movements of soils and rocks.

This chapter is the first of two that study the **geomorphology** of the Earth (i.e. processes and patterns that shape the landscape) and which examine slow, intermediate and fast processes of landscape change. In this chapter we start by covering tectonic processes which build and destroy landscapes on a large scale. The chapter then focuses on weathering and erosion which may seem small-scale but can produce an overall effect which is global. While weathering, erosion and deposition can also build landscapes (e.g. sand dunes at the coast or in deserts), the overall effect on the continents in total is that tectonic processes build landscapes and weathering and erosion including the action of rivers, glaciers and ice sheets (which are dealt with in Chapter 4), sculpt and can ultimately destroy landscapes. The environment that we see today represents one point in the overall evolution of the landscape which changes through time. The chapter has a specific section on soils, which are vital to our lives. Soils are a product of weathering and biological inputs and they slowly accumulate through time. Soils are also subject to further weathering and erosion as part of the larger-scale cycles of the Earth, but these processes can be affected by human action.

TECTONICS: CONTINENTS AND OCEANS

Plate tectonic theory

The Earth is roughly spherical although it is 42 kilometres shorter if you travel around the world via the poles compared to travelling around the equator. The 1,200 kilometres thick, inner core of the Earth is solid, hot (3,000°C) iron. The inner core is surrounded by a 2,300 kilometres thick layer of liquid iron-rich material forming the outer core (Figure 3.1). The next layer, as we move out from the centre of the Earth, is called the mantle (2,900 kilometres thick). The outer part of the mantle (180 kilometres) along with the overlying crust is called the lithosphere and this is rigid and floats on the more mobile asthenosphere. When rocks in the mantle are subjected to different pressures and heat they flow slowly. Large continents form a cold and rigid crust mainly made from granite rock around 35 to 70 kilometres thick. However, the rock on the ocean floors is made from basalt and forms a thinner, 6 to 10 kilometres thick, crust. Importantly (as we will see later), the density of the rock on the ocean floors is slightly greater than that of the continents.

The proof for plate tectonic theory has surprisingly only emerged in the last half century. Earlier explorers had noted that the shapes of the continents seemed like they would fit together (e.g. South America and Africa) and there had been scientists who studied fossil evidence during the nineteenth and early twentieth centuries such as Alfred Wegener who had suggested that all of the continents were once joined together in a single land mass called Pangaea, which slowly broke and drifted apart. However, it was only through surveys of the ocean floor, as part of naval submarine and nuclear research during the 1950s and 1960s that data was produced that showed how the continents really did move slowly around the Earth and that the oceans floors spread from their centres.

Detailed maps of the ocean floor showed that there are large mountain ranges running through the centre of the world's major

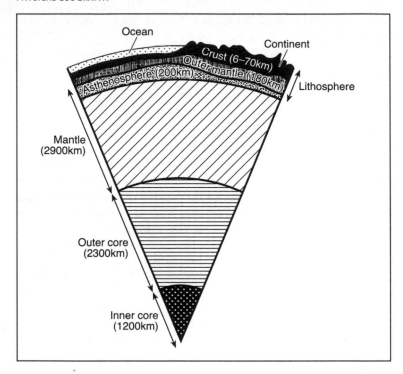

Figure 3.1 The interior of the Earth.

Note The schematic diagram is not drawn to scale.

oceans and that there are valleys in the middle of these mountain ranges. It was also found that the deepest parts of the oceans are located very close to the edge of the ocean rather than in the middle. The ocean floor was found to have 'magnetic stripes'. It is known that the Earth's magnetic field reverses every few hundred thousand years and the direction of the poles is recorded at the time when volcanic lava forms and cools. Alternating north or south-facing magnetic stripes occur right across the oceans and are orientated parallel to the mid-ocean mountain ridges showing that the ocean floor had formed at the mid-ocean and then moved slowly away on either side of the mid-ocean towards the continents. This provided the first real evidence that large masses of rock on Earth can slowly drift.

Within the mantle, there are massive, hot currents of liquid kept molten by radioactivity within the Earth. These currents form circulation cells within the mantle, a bit like the large atmospheric convection cells described in Chapter 1. When the rising currents reach the stiff lithosphere, they drag it along. In places the hot, rising mantle material can force its way through the crust of the sea floor. Here underwater volcanoes develop, forming a deep-sea mountain range. The lava from these volcanoes forms new crust as it cools. As the descending parts of the slow convection cell in the mantle separate, they can slowly (1–10 cm per year) drag overlying ocean crust with them (Figure 3.2). The great, deep trenches around the edges of the oceans are zones where old crust is forced down and melts into the Earth's mantle (Figure 3.2). The thickest (and oldest) rocks on the ocean floor are also found nearest to the continents. The oldest rocks on the ocean floor have been dated as 208 million years old. Therefore the ocean floor all over the world is young compared to the age of the continents where the rock is often between two to four billion years old.

There are several stiff plates moving across the Earth's surface which grind or rub against each other (Figure 3.3). Earthquakes occur most at the boundaries of these plates. The continents are

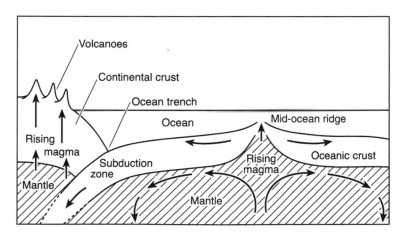

Figure 3.2 Mid-ocean ridge, sea floor spreading and ocean trench subduction zones.

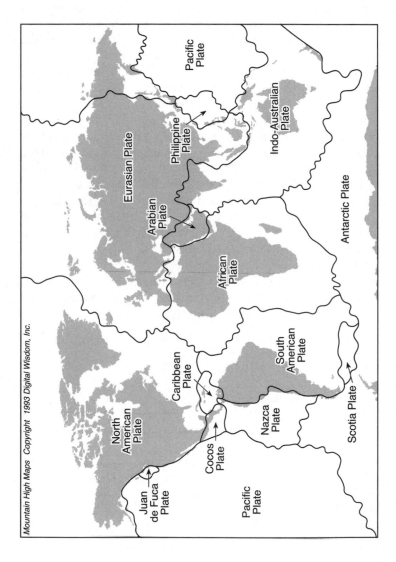

Figure 3.3 The main tectonic plates.

quite passive features of these moving plates since they just ride on top of them and, unlike ocean floors, they are not consumed into the mantle. Earthquakes occur because while plates slowly move you can imagine them having rough or bumpy sticking points. Over time an enormous force builds up and eventually the plates move in a jolt which is experienced as an earthquake.

Movements at the boundary between two plates can explain the nature of landforms found in these areas. Where plates are moving apart there are **divergent plate boundaries** (e.g. at the mid-ocean ridge) where new crust is formed. The lava formed at mid-ocean ridges is hot and very runny, forming gently sloping shield volcanoes. Volcanic eruptions with this type of lava (e.g. on Iceland which straddles a mid-ocean ridge) tends not to be explosive because gas bubbles can easily escape through the runny liquid, although occasionally large gas bubbles do emerge creating a scene with runny lava flying into the air. Eruptions may be in the form of walls of molten lava issuing from a linear crack in the Earth. Most divergent boundaries are in the mid-ocean but there are some within continents. The Syrian-African Rift Valley is a good example of a divergent plate boundary on land. As the valley has continued to deepen, it is now below sea level and some of it has filled with water (e.g. the Dead Sea is 339 metres below sea level).

Transform faults occur where plates slide past one another (e.g. San Andreas Fault, California). Here there is often little creation or destruction of lithosphere and there are few volcanoes at transform boundaries. However, these boundaries can be associated with frequent major and destructive earthquakes. The rates of movement can be from a few centimetres in a small earthquake to two metres in a large event.

When two plates move towards one another, at a **convergent plate boundary**, major physical features are formed. If one of the plates slides beneath the other, a **subduction zone** is formed. This happens where two ocean crusts collide or where denser ocean crust meets less dense continental crust and ocean crust then becomes part of the mantle at this point. This is why ocean crust is relatively young by geological timescales. Often this process also creates mountain belts as the crust thickens at the subduction zone. For example, the Nazca plate collides with the South American plate and is subducted below it creating the Andes Mountains and many volcanoes. The major destructive earthquake resulting in the **tsunami** that struck Indonesia and the eastern shore of the Indian Ocean on 26 December 2004, causing massive loss of life, occurred at a subduction zone.

Volcanoes produced around subduction zones can be very explosive and destructive. The oceanic crust is heated as it is carried down into the mantle. Water and other materials which are carried down with the plate are released, producing a mixture that rises to the surface. If the uppermost plate is oceanic, then basaltic volcanoes are produced which form into arcs of islands. When an oceanic plate collides directly with continental crust the oceanic plate moves under the continent. Water becomes trapped and causes the basaltic rock to melt under pressure. Rising magma starts to melt the continental crust of the overlying plate. This magma is very sticky and can result in a destructive volcanic explosion destroying large areas and killing many people. Examples of such volcanoes include Krakatoa (Indonesia), Vesuvius (Italy), Fujiyama (Japan) and Mount Saint Helens (USA). Sticky, slow-moving lava builds within these steep-sided volcanoes and once the lava ceases to flow, it cools producing a plug allowing a considerable pressure to build up within the volcano ready for the next eruption.

There are around 600 active volcanoes above the land or ocean surface (but tens of thousands on the ocean floor). On average around 50 surface volcanoes erupt per year although we tend only to hear about those that cause major destruction or disruption to travel. Volcanoes form at the end of a central tube or vent that rises from the upper mantle. There is often a crater which is a surface depression at the top of the volcano. Magma within the volcano slowly rises, building up pressure until conditions are right for eruption. Often the heat from the magma can also boil water in the ground resulting in hot springs and **geysers**. Volcanoes do not always form just at plate boundaries and more information on this is provided in Box 3.1.

Convergent boundaries compress rocks and deform them. This causes the rock to fold and rumple as if it were a piece of cloth being pushed together from both ends. This can produce mountain ranges which look like ripples when viewed from a plane high above. The most intense mountain building occurs when two continents collide. This is because they push into each other and one is

Box 3.1 Hot spots

There are some volcanoes that are not located at plate boundaries. These are at hot spots. They are probably located at the top of hot plumes within the mantle (a bit like an ocean or atmospheric circulation eddy). There are over 40 hot spots and many of these have formed in the middle of plates. For example, at Yellowstone National Park in the United States there is a hotspot which is responsible for amazing geysers, **mud pots** and other volcanic features. There is evidence that there was a massive volcanic eruption there 600,000 years ago depositing ash 12 metres thick up to 1,200 kilometres away. Calculations suggest that a 'supervolcano' may be due to erupt soon. If it does so then we will no longer be worrying about climate change but instead how to deal with a much larger and dramatic environmental catastrophe.

Where hot spots occur under the ocean and there is a volcano then, as the plate moves over the hot spot, a series of volcanic islands or seamounts (underwater mountains that have not reached the water surface) are formed. The best example of such a system is the Hawaiian island chain. The island of Hawaii is furthest to the east and is the site of the most active volcano on Earth at present. In fact the volcano, Mauna Loa, is the tallest structure on Earth when measured from its base on the ocean floor, at 10 kilometres tall. Mauna Loa and the whole island of Hawaii has only taken 1 million years to form which is very short on geological timescales. Then there are a series of islands, Maui, Molokai, Oahu and Kauai, in a line stretching more than 500 kilometres to the west. Each island is made up of an extinct volcano, with the volcanoes getting progressively older as you travel west. Further west the islands are so old and eroded that they disappear below the ocean water surface to become seamounts. This also happens because the westerly moving ocean floor becomes deeper towards the ocean trench.

not subducted below the other. This forces land to rise thereby creating a large mountain belt such as the Alps where Italy has moved north into Europe, or the Himalayas where India has collided into Asia. The Himalayan continental collision zone seems to have shortened the length of continental crust by 1,000 kilometres. Therefore, here the crust is thickened and compressed and as a result, the rocks are folded and deformed, crumpled and faulted.

The thickened crust sticks up like an iceberg floating on top of the mantle with a deep root. All ten of the Earth's highest peaks are found in the Himalayas. As the high mountains are eroded and mass removed then the floating root at the bottom moves up and exposes more rocks and minerals which have been altered by high temperatures and pressures.

WEATHERING AND EROSION

Weathering is the physical breakdown of rock whereas erosion is the transportation of the weathered material. Mountains that are built by tectonic processes are eventually worn down by weathering and erosion. The sediments produced are moved around often over vast distances by water, ice or wind and can be reincorporated into rock formation again over long timescales. Thus the Earth's surface is in a constant state of change. The rates of change vary with rock type, climate, slope conditions, ice and vegetation cover.

Types of rock

There are three main types of rock found around the Earth's surface: **igneous**, **sedimentary** and **metamorphic rocks**. Temperature at the time of formation, the mix of minerals present and pressure all interact to create varieties of these main rock types. Igneous rocks are formed when molten lava cools and hardens. If the molten rock is from a volcano, then the subsequent cooled and hardened basalt rock has small crystals. If the rock is able to cool slowly then larger crystals may grow producing a coarse grained rock such as granite.

Sedimentary rocks are produced by weathering of rocks followed by subsequent erosion and deposition of material. The deposited sediment can accumulate and eventually build up before being compacted and hardened over long time periods by the weight and pressure of sediments above and internal chemical changes. Rocks such as sandstone, siltstone or shales are good examples. These rocks often contain a record of the physical conditions present when the rocks were deposited, including fossils. In fact some rocks such as chalk or coal are almost entirely made from the remains of animals and plants.

Metamorphic rocks form through the partial melting and reforming of existing sedimentary or igneous rocks often under high pressure. Limestone and shale change to marble and slate when metamorphosed, for example. These rocks tend to be more resistant to weathering as they are harder than other rock types.

The rock cycle means that all rock types can convert into other types. All rocks can be melted and cooled to form igneous rock. All rock types can be weathered and eroded to form the layers of sediment that can eventually become sedimentary rocks. Under pressure and heat, igneous and sedimentary rocks can transform to metamorphic rocks.

Weathering

Weathering is the breakdown of rocks by physical and chemical processes. These processes often work together and biological processes can include both of these mechanisms. Some types of rock are more difficult to break down than other types of rock and so weathering can result in interesting landforms such as headlands or areas of protruding rock in an otherwise flat landscape (e.g. Ayers Rock – now known by its Aborigine name Uluru – central Australia) where one type of rock wears down more quickly than another in the vicinity.

Physical weathering

Physical weathering transforms rock by breaking it into smaller fragments through mechanisms such as freeze-thaw, salt weathering and thermal cracking. However, physical biological action through roots forcing openings in rock can be important locally. Freeze-thaw is the process by which water freezes in small cracks and expands by 9 per cent as it does so. This then forces cracks open further, eventually splitting the rock. The process is more active where temperature frequently fluctuates above and below 0°C.

Salt weathering is where salts within the environment form crystals in small cracks under desert conditions. The relatively bare rock surfaces in deserts along with large diurnal temperature ranges, and the excess of evaporation over precipitation, can lead to salts becoming concentrated in surface locations and cracking or flaking off the surface rock. Crystallisation occurs when temperature increases lead to the growth of salt crystals. Moisture inputs then cause the salt volume to increase and thermal expansion also occurs to the salts upon warming. This type of weathering is more common where coastal fogs can bring sea salts into desert areas such as Namib desert in Namibia.

Thermal weathering may also be important in deserts where daily temperature ranges may be high. This leads to the regular expansion and contraction of rock. Different minerals expand by different amounts when warmed. This creates internal stresses that weaken rocks and loosen particles. Rocks can often crack if you light a fire around them due to these stresses upon expansion.

Chemical weathering

Water acts as a solvent to dissolve rock. Rocks are made up of **bases** (calcium, magnesium, sodium and potassium), silica and **sesquioxides** (mainly with aluminium). Silica, is at least ten times less soluble than bases, but is at least ten times more soluble than the sesquioxides. Therefore, chemical weathering reduces the proportion of bases the most, followed by the proportion of silica. As material is exposed close to the surface, the atmosphere is able to assist weathering of minerals. The gases of the atmosphere such as oxygen, water vapour and carbon dioxide aid weathering (e.g. iron and oxygen can produce iron oxide (rust)). Small amounts of carbon dioxide from the atmosphere dissolved in rainwater make a weak carbonic acid which acts to weather rock. More intense rainfall combined with warmer temperatures and higher carbon dioxide concentrations may increase chemical weathering rates under climate change. However, when the carbonic acid reacts with the rock in the weathering process it produces other dissolved chemicals which then get transported away in solution. This can result in loss of the carbon dioxide from the atmosphere into oceans via river channels and could potentially act as a negative feedback on climate change.

Roots from vegetation put organic acids into the soil which help them to extract the nutrients needed by the plant and this also weathers soil. Many of the small animals in the soil (e.g. earthworms) pass soil material through their bodies, altering it both biochemically and mechanically as they extract nutrients from it.

Erosion

The removal of material in dissolved or particulate form can occur through several processes. Dissolved removal occurs in water. Water carrying the dissolved material removes it from the slope through the ground or over the surface. The concentration of solutes is generally highest in dry climates, but the total amounts of the dissolved material removed is less than in wetter areas. Once dissolved material is removed it generally travels far downstream. For some limestones, 90 per cent of their original volume can be dissolved, often producing dramatic landforms and caves known as **karst** landforms. Rapid karst development in wet tropical areas can result in amazing rock towers. Hot desert areas have the slowest karst development due to a lack of water for solutional weathering. Around 15 per cent of the Earth's surface has some karst landforms.

Soil material being transported downslope can do so as a large mass or as independent particles. During mass movements (which can be very fast, or very slow), a section of rock or soil moves together. Where large flows of water generate mass movement of sediment this tends to be faster than drier mass movements of sediment. The net effect of all forces operating on a mass of material controls when the material will move. Forces that encourage movement include gravity, water and wind. Flowing water can detach fragments of rock or soil if it passes over them rapidly by picking up material from the surface, or by detachment of soil grains by raindrop impact. Friction and **cohesion** resist movement. Material begins to move when the forces promoting its movement become larger than the resistance forces. The **safety factor** is the ratio of these forces. In simple terms, moving material will slow down and stop when it meets lower gradients or where water carrying the material spreads out and moves more slowly, or seeps into the ground.

There are many types of mass movements. In rapid mass movements, there are slides where the mass moves as a block, and flows within which different parts of the material move over each other at different speeds. Fractures in rock provide weaknesses which fail and slabs of rock can slide off downslope. Toppling can also occur when columns of rock become overhanging. Flows occur when there is more water mixed into the moving mass, in proportion to the amount of sediment. In a slide there is little water within the moving material itself although water may have helped overcome friction to initiate the event.

There are also slow mass movements. As a whole unit, a mass of rock or soil moves slowly downslope. The slow mass movement of soil is called soil creep and typically operates at 1 to 5 millimetres per year. The movements are caused by expansion or contraction heaving (e.g. wetting and drying of soil or freeze-thaw action). Other movements are usually caused by biological activity which mixes the soil in all directions. When this occurs on sloping ground then gravity will result in more downhill movement than uphill and there is a gradual transport of material. Depending on the environment, one of the processes of creep may be more domi-nant. In cold, upland areas freeze-thaw is probably the most important, whereas in tropical forests biological mixing may dominate.

Box 3.2 Plough creep

Humans can accelerate soil creep through ploughing and this is often known as tillage erosion. Each time the soil is turned over by the plough the soil moves. If the ploughing is directly up or downslope then there tends to be an overall downhill movement of soil when the soil settles back in the period after the ploughing. Where the ploughing direction follows the contour and cuts across the slope then material is moved either down or upslope depending on the direction of the plough. Where the plough turns soil downhill then this will result in an overall movement approximately a thousand times greater than soil creep. Contour ploughing in both directions as the plough moves one way across the slope and then the other still produces movement a hundred times greater than natural creep. Over the last few hundred years tillage erosion may have been responsible for more soil movement in many areas than natural soil creep has over the last 10,000 years.

Water moves particles in what are called 'wash processes'. Rain-splash, rainwash and rillwash are the most important of these wash processes. The impact of raindrops can detach material which then jumps into the air. The splash can cause the sediment to move either up or downslope but because of gravity there is an overall downslope movement in total. The rate at which rainsplash transports material is similar to the rate of transport by soil creep. However, while soil creep occurs over a depth of soil and transports a large mass of material, rainsplash only functions on the surface, moving individual particles. Ensuring there is a good vegetation cover to protect the soil surface from rainsplash forces is a good way of reducing soil movement by rainsplash.

If raindrops land on flowing water moving over the land surface then their direct impact on the soil surface is reduced. However, the flowing water itself can carry material. Where the flow over the surface is shallow, the combined effects of raindrop impact, which detaches sediment, and transport by flowing water over the surface, are very effective and this combined process is called rainwash. When the water depth over the surface is deeper than 6 millimetres, raindrop detachment is weak and so the initial movement of a particle is more related to flowing water in a process called rillwash. This erosion process is common in major storms. Many poorly vegetated areas develop temporary rills which are channels formed during storms. Wetting and drying, or freeze-thaw accumulates material that infills the rills between storms. However, in a large storm, channels may form which are too large to be refilled before the next event and these are known as **gullies**. These channels collect water in subsequent events, rapidly enlarging the gullies further.

Where there is lots of weathered material but the erosion process cannot carry it far (e.g. rain splash hitting the surface and moving particles) then the erosion process is transport limited. In situations where there is not much sediment available to be transported but the transport processes could, if there was more sediment available, carry a lot more, then this is known as a supply-limited condition. Places where such transport-limited conditions operate have a tendency to have a good cover of vegetation and soil, and over time the steepness of slopes often declines. Landscapes where removal of material is mainly supply-limited tend to have little vegetation and soil and have steep slopes which remain steep as they erode. The landscape can therefore contain many clues to the processes forming it. Convex hillslopes are associated with creep or rainsplash. Concave profiles are generally associated with rillwash. Mass movements generally lead to slopes with quite a uniform gradient except in situations of exceptional activity (e.g. coastal cliff erosion). Thus, landslides produce a landscape with uniform slopes. Semi-arid areas have stony, shallow soils with little vegetation; here hillslopes tend to be concave. Temperate wet areas are generally dominated by creep, mass movements and solutional erosion under a dense vegetation cover and deep soils; here hill-slopes tend to be convex. You can also examine smaller features to see what processes have been operating, for example by checking whether there are small mounds of sediment behind clumps of vegetation as an indicator of active wash processes.

The wind can be an effective agent of sediment transport if the right type of sediment is available. Wind-blown sediment transport is called **aeolian transport**. Aeolian transport dominates in arid and semi-arid environments where there is little water. Strong winds are needed just to carry small particles. At typical wind speeds, medium sized sand grains (up to 0.5 millimetres) are the largest grains that can be transported. However, the wind can carry finer dust thousands of kilometres. The amount of material removed from the Sahara by wind transport is estimated to be around 260 million tonnes per year. Dune deposits are familiar landforms produced by wind. However, erosional features in rock can also be formed because sand and dust particles in the air can be abrasive. Abrasion by sand is usually close to the ground (within 2 metres) as sand cannot be carried very high due to its large size. However, finer particles can also abrade at higher levels, smoothing rock surfaces and even whole hills, which are often known as **yardangs**.

SOILS

Soil composition and formation

The weathering processes described above result in rocks becoming fragmented providing small habitats for plants which then add organic matter. This helps form soil. Soil is of crucial importance to humans. It acts as a zone for plant growth, which provides crops, supports animal life and holds water, influencing the amount and quality of water in rivers and lakes.

Soil is made up of minerals, organic matter, water and air. The amount of each of these components influences the properties of a soil. In most soils, the majority of the solid material is mineral matter which has been derived from rock weathering. Often only 2 to 6 per cent of the soil is organic matter but it is still very important. Soil organic matter consists of decomposing plant and animal debris known as **litter**. It also consists of organic matter more resistant to decay known as **humus**, and living organisms and plant roots known as the soil **biomass**. Indeed, soil usually contains billions of bacteria in every handful. Litter is decomposed by soil organisms to produce humus which is a stable end product, resistant to further decomposition. Plant nutrients, especially nitrogen, phosphorus and sulphur are released as litter breaks down and this process is known as **mineralisation**. Soil organic matter holds mineral particles together, stabilising the soil; improves the water-holding capacity; improves aeration; and it is a major source of nutrients, important for soil fertility.

Air and water fill the gaps between solid soil particles. Soil air is very important as soil animals, plant roots and most microorganisms use oxygen and release carbon dioxide when they respire (breathe). In order to allow soil organisms to survive, oxygen needs to move into the soil and carbon dioxide must be able to move out of the soil. Therefore, the aeration of the soil is an important component influencing biological activity and litter decomposition.

Soil water contains dissolved substances which are important for uptake by plant roots and moves dissolved chemicals through the soil (both up and down) making them available for plants. Interestingly, water is held in the soil despite gravity forces pulling it down. Even in very hot, dry deserts, water is still found in the soil showing that the forces holding water in the soil must be strong. This water remains in the soil because the combined chemical attraction of the water molecules to each other and the attraction of water to the soil particles is greater than the gravitational force. If you dip the end of a tissue into a bowl of water, then you can watch the water being drawn up the tissue showing that water does not always flow downhill. This is capillary action. Smaller pores exert stronger attraction forces on the water than larger pores. Therefore, capillary water will be driven to move from wetter parts of the soil to drier parts because the drier parts have more small pores that have little water thereby exerting a capillary attraction force on the water. This capillary action is also how plant roots draw water into the plant. If the soil is coarse, generally consisting of lots of large particles and large pore spaces between the particles, then it will not be able to hold as much water as a finer soil with smaller particles and smaller pore spaces. This explains why sandy soils cannot hold much water and are not as good at supporting plant growth as finer textured soils which have more small pores.

Soil formation takes place over thousands of years. The main input of soil material comes from the weathered rock below the soil. Mineral particles are released by weathering and contribute to the lower layers of the soil. Surface accumulation of organic matter from plants and animals is also important, as is dissolved material in water and particles carried by precipitation and the wind. The main losses of material from soils occur through wind and water erosion, plant uptake (but this usually gets returned to the soil after death if the system does not have crop material removed from site) and **leaching**. Leaching is the removal of dissolved soil material. The leaching process is most rapid where there are large water inputs to the surface and where the soils are well drained (e.g. in an irrigated agricultural field with coarse soils with under-drainage installed). The percolating water carries dissolved substances downwards, depositing some in lower layers but some of the dissolved material may be completely washed out of the soil.

The factors that influence soil formation include climate, the 'parent material' (i.e. the weathered rock matter), the slope and organisms. The most influential factor is climate since it determines the moisture and temperature conditions for soil development; maps of major soil types often follow the climate zones. Soils in high latitudes are often very shallow and develop slowly while soils that are several metres deep are typical of tropical areas. Parent material influences soil formation through the influence of weathered material on soil processes while slope steepness, aspect and altitude all affect the local climate as well as drainage and erosion conditions. The type of vegetation influences the type and amount of litter that is returned to the soil while different soil types support different vegetation communities. In conifer forests there will be a deep litter layer of thin waxy needles which only decompose slowly. Vegetation also protects the soil from water and wind erosion by intercepting rainfall, decreasing the role of rainsplash (see section on erosion above).

Soils are often classified by the nature of their vertical profile. If you dig a hole through a typical soil, from the surface to the bedrock, the soil will be made up of a series of horizontal layers known as **soil horizons**. In some soils, the horizons have very distinctive colours and clear boundaries while in others the changes with depth may be gradual with rather uncertain boundaries. Soil horizons are given letters according to their formation and their relative position in the profile. The uppermost O horizon is dominated by an accumulation of fresh and partially decomposing organic matter. Below this the dark A horizon contains humus and minerals. The typically pale E horizon is dominated by leaching or removal of material by plant roots while the underlying B horizon is often a zone of accumulation of matter from leaching above and weathering below. There is usually a transition from the B horizon into the C horizon which is mainly weathered parent material known as the **regolith**. The bedrock below this is typically designated as the R horizon. Not all soils contain all of the layers described above.

'**Podzolisation**' (producing soils called podzols) may occur where there is strong leaching often where there is plentiful rainfall and good drainage, typically under forests or heaths. Organic acids from litter are washed through the soil and react with iron and aluminium compounds that are transported downwards from the E horizon by percolating water and deposited in the B horizon. Podzols tend to be rather unproductive for agriculture as fertilisers are readily washed away and they are acidic. In waterlogged conditions there are reactions with iron products in the soil driven by microorganisms in a process called **gleying**. The soil becomes grey or bluish. **Laterisation** (ferralitisation) occurs in tropical soils where high temperatures and plentiful rainfall result in fast rates of weathering and leaching so that there are horizons depleted in bases (e.g. calcium, magnesium, potassium and sodium) and enriched in silica and oxides of aluminium and iron. In arid and semi-arid areas, water is drawn to the soil surface and as the water evaporates salts are left behind at or near the surface.

There are several different soil classification systems and each system contains around 15 to 30 different soil types, often with subcategories for main types. Often the words used in one classification are different for the same soil in another classification scheme which causes confusion. While there are too many examples to describe in this book some names from the US Department of Agriculture system are: vertisols (swelling clay soils with deep, wide cracks), histosols (organic rich soils), andisols (formed on volcanic parent material, especially ash) and oxisols (red/yellow/grey soils of tropical and subtropical regions with strongly weathered horizons enriched in silica, clay and oxides of aluminium and iron and are acidic with low nutrient status).

Physical properties of soils

Soil texture and structure influence how soils work and how they can be managed. Soil texture refers to the relative proportions of sand-, silt- and clay-sized materials in a soil. Clay particles are smaller than 2 micrometres (two-millionths of a metre), silt is between 2 and 60 micrometres while sand is between 60 and 2,000 micrometres in diameter. The texture controls the water holding capacity, aeration, drainage rate, organic matter decomposition rates, compaction, susceptibility to water erosion, ability to hold nutrients and leaching of pollutants. Figure 3.4 shows how soil texture classifications are based on the relative proportion of each particle size. For example, reading the figure shows that if a soil is 40 per cent sand, 30 per cent silt and 30 per cent clay it would be classified as a clay loam.

Good soil structure is important for achieving well-drained and aerated soils. Soil structure refers to the arrangement of soil particles. Soil particles usually attach themselves to each other in formations called **peds**. Where a soil lacks peds, such as in a sand dune, the soil is described as structureless. Soil structure is characterised in terms of the shape, size and distinctness of these peds with four principal types: blocky (roughly equally sized on each side, almost cube-shaped, but the peds can be angular or more rounded), spheroidal (sphere-shaped), platy (horizontal plates) and prismatic (vertically elongated columns of soil with flat face). Clay particles and organic compounds largely hold peds together. As a result, coarse-textured soils tend to have weakly developed structures, whereas fine-textured soils generally have moderate to strong structures. The strength of peds influences resistance to erosion and the ease of cultivation.

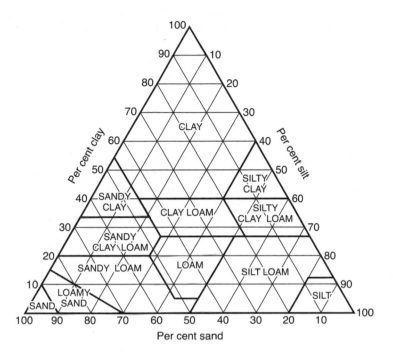

Figure 3.4 A common classification for soil texture.

Chemical properties of soils

Soil chemical properties are strongly influenced by parent material and organic matter content which provides an amount of tiny clay and organic particles. Clay minerals are formed from the weathering products of aluminium and silicate minerals. As clay minerals are small, a volume of clay will have a large surface area around all of its particles in total compared to the same volume of sand. To test this out, measure the surface area of three soccer balls that fill a box and then measure the surface area of the number of tennis or golf balls it takes to fill the same box. An ion is an atom or group of atoms with an electrical charge (either positive or negative). Clay particles have a negative electrical charge so these ions can attract and hold water and cations (positively charged particles). Therefore, they have a fundamental influence on both the physical and chemical properties of the soil. The concept of the **cation exchange capacity** is an important one. It is essentially a measure of the soil's ability to hold and release various elements such as plant nutrients. Experiments during the nineteenth century showed that if you add ammonium chloride (as part of nitrogen fertiliser) to the top of a soil a solution of calcium chloride came out of the bottom. The cations of ammonium and calcium were rapidly exchanged in this process. The process is also reversible. The negative charge on the clay and organic humus particles is balanced by positively charged cations that are attracted to the clay and humus particles. These cations are referred to as exchangeable cations because cations in the soil solution can displace the cations on the clay surface. This exchange between a cation in soil water solution and another on the surface of a clay particle is cation exchange. Cation exchanges are balanced reactions so that if an ion with two positive charges such as calcium (Ca^{2+}) is washed through by a solution of sodium (which has a single positive charge; Na^+) then it will take two sodium ions to replace the one calcium ion. The cation exchange capacity is the ability of an amount of soil to hold cations and this depends on the overall negative charge of the clay particles present. The cation exchange property controls fertility and acidity and also means that soils act as an important buffer between the atmosphere and groundwater thereby potentially reducing pollution of water courses.

The acidity of a soil is important since it affects many soil processes, the plants that grow and what happens to some pollutants. Many polluting heavy metals become more soluble in water under acid conditions and can then move downwards with water through the soil to groundwater or river water. The concentration of hydrogen ions in soil solution determines whether a soil is acidic, neutral or alkaline. Concentrations of these ions are very small and so the pH system was developed. Low numbers (starting at 1) on the pH scale are acidic, 7 is neutral and large numbers up to 14 are alkaline. In the pH scale a change of one unit represents a 10–fold change in hydrogen concentration. So a pH of 5 means the soil solution has 10 times the concentration of hydrogen ions than at a pH of 6. Most soils have a pH of between 3.5 and 9 and very low values are often associated with soils rich in organic matter, such as peat.

Green plants cannot grow properly without 16 essential elements available in the correct proportions. The availability of these essential nutrients for plants is influenced by soil pH. On the basis of their concentration in plants the 16 elements are divided into macronutrients (carbon, oxygen, hydrogen, nitrogen, phosphorus, sulphur, calcium, magnesium, potassium and chloride) and micronutrients (iron, manganese, zinc, copper, boron and molybdenum). As a comparison, the typical amount of potassium in a soil is 1.5 per cent of the total mass whereas molybdenum only accounts for a hundred thousandth of a percentage point. A pH range of six to seven is generally best for plant growth as most plant nutrients are readily available in this range. High soil pH results in phosphorus and boron becoming insoluble and unavailable to plants. Most nutrients are more soluble in low pH (acidic) soils, which can result in high or toxic concentrations of them. On the other hand phosphorus and molybdenum become insoluble at low pH and are unavailable to plants.

Humans and soil

Human activity can alter the soil. It is thought that the total global area of soil degraded by humans (over 20 million square kilometres) exceeds the amount currently being used for farmland. This degradation is due to deforestation, overgrazing and poor agricultural management. Degradation includes soil erosion, soil acidification, soil pollution, the reduction of organic matter content and **salinisation**.

Over the last half century it is thought that between one-third and a half of the world's arable land has been lost by erosion. While erosion is a natural process, humans have dramatically accelerated this process. Farmland is often left bare or has a low vegetation cover for a considerable period of the year and old obstacles to erosion like large woody plants have often been removed. This makes it easier for the wind or wash processes to transport surface sediment. Tractor wheels also compact the soil making excellent fast routes for water flow over the surface often accelerating erosion. Accelerated soil erosion results in a reduction of soil depth because soil is lost at a faster rate than it is being formed, often with the more organic rich upper layers being lost first. The erosion can sometimes lead to large gullies forming, making it difficult to farm. The sediment lost through erosion moves downstream and can often silt-up roads, reservoirs and water courses. Some agricultural chemicals are also bound to the sediment and so this pollutes water courses. Erosion control often involves reducing grazing intensity, planting strips of vegetation to form barriers to reduce the wind and catch sediment, use of cover crops when the surface would normally be left bare, more careful ploughing and building terraces to reduce the slope and catch any sediment (so the landscape looks like a series of steps coming downhill).

Soil pH has been reduced by the burning of fossil fuels which has resulted in rainwater becoming more acidic. Also the harvesting of crops and overuse use of nitrogen fertilisers causes acidification. Soil acidification increases the solubility of heavy metals in the soil which can be toxic to plants, reducing growth rates or altering the types of plants for which the soil is suitable (e.g. forest decline in central Europe). The soil organisms may also be affected with species changing towards those which are more tolerant of acidic conditions, resulting in a slower rate of litter decomposition.

Pollutants such as heavy metals, pesticides and fertilisers can damage soil and may result in water contamination. Heavy metals (metallic elements with a density greater than six grams per cubic centimetre) such as copper, lead, zinc and mercury are present naturally in soil, but atmospheric pollution and application of sewage sludge, farm wastes and leaching from landfill sites can add heavy metals to soil. The worst areas for heavy metal pollution are those around industrialised regions such as in north-west Europe. Mining, smelting, energy generation, agriculture and the wear of vehicles and other machinery are sources of heavy metal pollution to soils. Heavy metals build up in soil as they become bound to organic matter and clay minerals and are generally not taken up by plants. However, if the soil becomes more acidic this can result in heavy metals being released into soil water and becoming available for plants to take up, or for leaching into rivers, lakes and groundwater. This is the crucial phase because if this happens then the crops we eat may have toxic levels of heavy metals, or the water becomes dangerous to drink.

Modern agriculture relies on pesticides for crop protection and disease control, and fertilisers to provide additional nutrients such as nitrogen, phosphorus and potassium, to maintain good plant productivity. Pesticides can often attack living things that they were not intended for and can leach into groundwater and rivers. Fertiliser use has increased tenfold over the last 50 years. This is fine as long as it has been applied in the correct amount for the soil and crop and at the right time of year so that it is not wasted and simply washed out of the soil. However, river water, groundwater and lakes have seen increased nitrate concentrations related to the leaching of fertilisers. Nitrates in drinking water are a health hazard and can damage river- and lake-based plants and animals. **Organic farming** relies on biological processes for crop and livestock production rather than using manufactured pesticides and fertilisers. Practices such as crop rotations, varieties more resistant to disease, use of plants that take nitrogen out of the atmosphere and application of compost and manures are common in organic farming. As yields tend to be lower than in conventional farming, food prices tend to be greater, but clearly there is a balancing act between food price and the wider price of environmental pollution through manufacture of agrochemicals or potential pollution from leaching.

Intensive agriculture has reduced soil organic matter. Soils with less than 1.7 per cent organic matter may be close to becoming a barren landscape rather like a desert. This is known as **desertification**. In the Mediterranean area 75 per cent of the landscape has a low (3.4 per cent) or very low (1.7 per cent) soil organic matter content and so desertification is a major cause for concern. Organic matter content has been reduced by the abandonment of crop rotation, ploughing up grasslands and the burning of stubble (vegetation remains after cropping). These all reduce the amount of organic matter being returned to the soil. Reversing these actions can allow some recovery of organic matter content (e.g. ploughing stubble back into the soil rather than burning it and increasing the proportion of grass cover).

Salinisation occurs when soluble salts of sodium, magnesium and calcium accumulate so that soil fertility is reduced. This is mainly a problem in warm, dry regions where evaporation and upward movement of water in soils exceeds downward movement from rainfall and percolation. Furthermore, irrigation of the land with water of a high salt content (i.e. where a lot has evaporated before being used so that the concentration of salts in the water is greater) worsens the situation. In some countries around 10 per cent of their arable land is affected by salinisation.

The above human impacts on soils have led to concerted efforts, legislation and policy to protect and restore soils as an important underpinning resource for human life. However, much more effort is needed particularly as the world's population continues to grow, placing more demand on our soil resource.

SUMMARY

- The Earth is 4.6 billion years old.
- The Earth's crust is formed of moving plates; at the edges of plates earthquakes and volcanic activity are common.
- As continents collide mountains form.
- In the centre of oceans new crust is formed; at the edge of oceans the crust sinks back into the mantle. Ocean crust is relatively young being less than 200 million years old whereas continental rocks can be billions of years old.
- Weathering by physical and chemical processes wears down rock. Climate and rock type are important controls of weathering process rates.
- Erosion transports weathered material by water, wind, and slow and fast mass movement.
- Soil is made up of minerals from weathered rock, organic matter, water and air.
- Soil formation is affected by climate, the parent material, topography and organisms.
- The texture of soil particles, structure and chemistry of a soil are crucial in determining its water and nutrient exchange capacity and hence its use for plant growth.
- Movement of materials vertically and laterally through the soil dissolved in water is an important process for transferring nutrients for plants and impacting stream water quality.
- Careful soil management is required as humans have degraded large areas of soil across the Earth through poor agricultural practice and pollution.

FURTHER READING

Brady, N.C. and Weil, R.R. (2007) *The Nature and Properties of Soils*, fourteenth edition, Upper Saddle River, NJ: Prentice Hall. This is a popular soils textbook; many of its examples are North American.

Bridge, J. and Demicco, R. (2008) *Earth Surface Processes, Landforms and Sediment Deposits*, Cambridge: Cambridge University Press. A long, detailed, but very well illustrated textbook popular among students.

Grotzinger, J. and Jordan, T.H. (2010) *Understanding Earth*, sixth edition, New York: W.H. Freeman and Company. This is a full colour text with additional media materials to explain the internal workings of the Earth and especially those around plate tectonics, earthquakes and volcanism.

Holden, J. (ed.) (2008) *An Introduction to Physical Geography and the Environment*, second edition, Harlow: Pearson Education. A textbook with expert contributors providing more in-depth material on all of the topics covered in this chapter.

Morgan, R.P.C. (2005) *Soil Erosion and Conservation*, third edition, Oxford: Blackwell. This is a student textbook with good examples.

Summerfield, M.A. (1991) *Global Geomorphology*, Harlow: Pearson Education. This is a detailed text for those who want to pursue geomorphological topics in a lot more depth. There is excellent use of illustrations and examples.

Yatsu, E. (1988) *The Nature of Weathering*, Tokyo: Sozosha. This is a very thorough treatment of the processes for those who want significantly more detail and equations.

4

WATER AND ICE

WATER

The water cycle

Water evaporates from oceans, soil, living things (transpiration), rivers and lakes. At some point this water vapour condenses to water or ice in the atmosphere and returns to the Earth's surface as precipitation. Some of this precipitation will infiltrate into the soils and rocks below the surface where it flows more slowly to river channels or sometimes directly to the oceans. The water remaining on the surface and in the upper layers of the soil will partly evaporate back to the atmosphere and partly run into lakes and rivers, many of which flow into the sea. There are four main stores of water. These are the world oceans, polar ice, terrestrial waters and atmospheric water. The oceans hold 93 per cent of the water, polar ice holds 2 per cent, the soils, lakes, rivers and groundwater hold 5 per cent and the atmosphere holds a thousandth of 1 per cent of water resources. The water held in glaciers and polar ice or deep within some rocks may be stored for several thousand years. The water held by plants may just be stored for a few hours.

Water movement through the landscape

Precipitation can hit the land surface or be intercepted by vegetation. Some intercepted water can evaporate while some can flow down plant stems or drip from leaves in order to reach the land surface. The water that reaches the land surface can either infiltrate into the soil or pond-up and flow over the surface as overland flow.

Infiltration is influenced by vegetation cover, soil texture, soil structure, the amount and connectivity of pore spaces and compaction. People often measure the **infiltration rate** of soils, which is the volume of water passing into the soil per unit area per unit time. They measure it because it is useful information to understand how fast water can soak into a soil and therefore how much water might run over the land surface during heavy rain. Water moving over the land surface tends to get into rivers more quickly than water that infiltrates into soil and moves within the soil. The maximum rate of infiltration when there is a plentiful water supply is the **infiltration capacity**. The infiltration capacity of a soil generally decreases during rainfall. Therefore, if rainfall is at a constant rate then the water arriving at the soil surface may at first all infiltrate into the soil but then as infiltration capacity decreases more of the water will run over the land's surface. Compaction of the soil, surface crusts or a frozen surface can restrict infiltration even if within the soil itself water percolation rates could be quite fast. These restrictions to infiltration can lead to fast generation of overland flow and potentially large river flood peaks. Soils with lots of humus and a deep litter layer, such as those within tropical rainforests, tend to have large infiltration capacities.

If surface water supply is greater than the rate of infiltration then overland flow can begin once small surface depressions begin to overflow. This process is called **infiltration-excess overland flow**. This type of flow is uncommon in many temperate areas except in urban locations, along roads and paths or perhaps along compacted soils in arable fields created by tractor wheels or overgrazing by animals. Infiltration-excess overland flow is more common in semi-arid regions where soil surface crusts have developed and rainfall rates can be rapid. It is also more likely in areas where the ground surface is often frozen, such as northern Canada or parts of Siberia.

However, one of the most common misconceptions is to think that infiltration-excess overland flow is the only way overland flow can be developed. In fact, there is another important process known as **saturation-excess overland flow**. When all the pore spaces are full of water then the soil is said to be saturated and the

water table (highest point below which the soil or rock is saturated) is at the surface. Water can therefore leave the soil and run out over the surface. This is common in shallow soils or at the bottom of hillslopes where water running through the soil will collect, keeping the pore spaces full of water for long periods allowing water to return to the surface at this point. This means that saturation–excess overland flow can occur long after it has stopped raining. If it is raining then this additional rainwater will find it difficult to enter the soil if it is saturated and so saturation–excess overland flow can be a mix of fresh rainwater and water that has been within the soil for some time.

The landscape area that produces saturation–excess overland flow varies through time. During wet seasons a larger area of soil will be saturated and able to generate saturation–excess overland flow than during dry seasons. If the **catchment** starts off relatively dry then during a rainfall event not much of the area will generate saturation–excess overland flow, but as rainfall continues then more of the catchment becomes saturated, especially in the valley bottoms, and therefore a larger area of the catchment will produce saturation–excess overland flow. This means that the source areas for overland flow are variable whereas the source areas for infiltration–excess overland flow tend to be the same.

Water moving through soils or rocks is called **throughflow**. Most rivers around the world are mainly fed by throughflow. This process can maintain river flow during dry periods (the **baseflow** of the river). There are different ways water can move through soil or rock and this affects how quickly water can get to a river channel and therefore the typical nature of the response of a river to rainfall. **Matrix flow** is where water moves through the small pore spaces within a rock or soil whereas in **macropore flow** water moves through larger cracks, fissures or continuous channels within the rock or soil thereby bypassing contact with most of the soil mass itself. It is possible to estimate the rate of water flowing through the matrix of a saturated soil or rock using techniques such as dye tracing, pumping water out of a well and timing the length of time it takes to rise up the well again, or taking small samples back to the laboratory and running water through them. Most water in soils tends to move through the matrix and therefore there are opportunities for cation exchange (see Chapter 3 on soil chemical properties) and changes to the chemistry of the water. However, some soils have lots of macropores within them, many of which actively transport water. At some sites measurements have suggested half of the water moving through soils could move through macropores. If an arable field has many macropores, surface fertiliser applications may get washed through the macropore channels and may not enter the main part of the soil. In this situation, the fertiliser will not be available for crop use and may enter the stream system leading to water quality problems. Macropores can be formed by soil animals, plant roots and cracking during dry conditions or from small landslips. Some macropores can be several metres in diameter, particularly in erodible soils such as in semi-arid south-east Spain or Arizona or in karst limestone cave systems. Turbulent flow within these large macropores in soils can rapidly enlarge them further until eventually they collapse to form gullies.

Groundwater

The term groundwater can be defined in many different ways. Some view it solely as water within rock. Another working definition is that groundwater is water held below the water table both in soils and rock. In any case water held within the ground is of worldwide importance. In many catchments water is supplied to the stream from groundwater in the bedrock. This is water that has percolated down through the overlying soil and entered the bedrock. Rock has small pores, fractures and fissures. Groundwater stores around 30 per cent of the Earth's freshwater (i.e. not salty). However, if this groundwater is to be available to supply river flow the rock or soil needs to be permeable, enabling water to flow through it. Layers of rock porous enough to store water and permeable enough to allow water to flow through them in economic quantities are called **aquifers**.

Some countries, such as Austria, Hungary and Denmark, almost entirely rely on groundwater. France obtains 56 per cent of its drinking water from groundwater while the UK obtains around 30 per cent from this source. The USA obtains around half of its water for all uses from groundwater with a quarter of drinking water coming from aquifer resources. However, a greater proportion of groundwater is served to communities with smaller populations which makes quality control more difficult (e.g. from a small well supplying 100 homes). People have obtained groundwater for thousands of years through digging wells or collecting water from springs which are zones of groundwater emergence. However, if the amount of water being withdrawn from the ground by humans is not replaced at the same rate from inputs to the aquifer then groundwater levels will fall. Indeed a rise and fall in water table naturally happens seasonally in many areas depending on rainfall and **evapotranspiration** rates. However, in many places groundwater has been greatly depleted, particularly where rates of replenishment are slow and the water that is being abstracted may be hundreds or even thousands of years old. The costs of pumping out deeper water are greater than water near the surface. Subsidence has resulted in some places due to groundwater depletion, such as in Mexico City, or Tucson, Arizona, USA (see Box 4.1). Because of the links between groundwater and baseflow, rivers can dry up if groundwater is over-abstracted. In coastal areas, where the aquifer is connected directly to the sea, groundwater abstraction can lead to the intrusion of salty water to replace the lost groundwater under land. This is a problem, for example, along lots of Australian coastlines where there are major cities, and on the shores of the Persian Gulf.

River flow

Water flowing through and across a land surface is called runoff (note that runoff does not just mean overland flow). Runoff often flows into rivers or lakes. The area of land that could potentially drain into a river or lake is known as the catchment area or **watershed**. Rivers can be fed by throughflow or by overland flow and the relative proportions of the different types of flow can determine how quickly the river flow changes during rainfall events or seasonally. River flow is crucial for aquatic life, water availability for reservoirs, abstraction for human use and flooding.

River flows can change during individual rainfall or melt events, or remain fairly stable, depending on the nature of the environment. There tends to be a lag between the precipitation occurring and the peak discharge of a river. This lag time is affected by the runoff processes discussed above. Where infiltration-excess over-

Box 4.1 Groundwater and subsidence in Tucson, USA

Tucson, Arizona lies in a semi-arid area surrounded by mountain ranges and desert. The population and its water use, including that for farming, grew rapidly at the start of the twentieth century and most water then had to come from groundwater wells as surface water supplies had been outgrown. Abstraction grew at a much faster rate than replenishment from rainfall or snowmelt from the mountains nearby. This meant water tables dropped in some places by 80 metres. The Santa Cruz River around which the city is based used to flow all year round but now only flows for short periods during the year as there is no baseflow into the river. During the mid-twentieth century hundreds of fissures, gullies and collapse features formed creating havoc for roads, pipelines and other infrastructure. In some places, subsidence of the general surface level occurred by around 8 metres. Water was therefore urgently needed from other sources to prevent further subsidence and also to ensure the city would have enough water to survive. A massive engineering project was devised which diverted water from the Colorado River for around 400 kilometres across the desert (and uphill) to towns across central Arizona including Tucson. The link was made in 1992 but the taste and odour of this new water, which had travelled across the desert and undergone evaporation and concentration of dissolved matter, was not to a lot of people's liking. Therefore, much of this new water was used instead to replenish groundwater, being filtered on its percolating journey before later being abstracted. Other activities such as recycling waste water are now being used to encourage the slow recharge of the aquifers in the Tucson area.

land flow dominates the runoff response then the **hydrograph** is likely to have a short lag time and high peak flow. The hydrograph (a graph of river discharge through time) will therefore be quite steep (perhaps even spiky) in shape, rising quickly from low flow to the peak flow. Urbanisation increases flood risk as it reduces the infiltration capacity of the surface through construction, leading to rapid runoff to the river resulting in steeply rising hydrographs with sharp peaks. If matrix throughflow dominates runoff response then the river may rise and fall very slowly in response to precipitation and the peak may be small. However, since throughflow contributes to saturation–excess overland flow then throughflow can still lead to rapid and large flood peaks. In some soils such as peat only a small amount of infiltration may be needed to cause the water table to rise to the surface. In other soils there may even be two river discharge peaks caused by one rainfall event. This might occur where the first peak is saturation-excess overland flow dominated and the second peak a little while later may be much longer and larger and caused by subsurface throughflow accumulating at the bottom of hillslopes and valley bottoms before entering the stream channel. Throughflow may also contribute directly to storm hydrographs by a mechanism called piston or displacement flow. This is where soil water at the bottom of a slope is rapidly pushed out of the soil by new fresh infiltrating water entering at the top of a slope.

Figure 4.1 shows flows in the form of hydrographs over one year for two nearby rivers where the climate is the same. Despite being in the same area the flows are very different between the rivers. The flows in the River Nol appear to be dominated by baseflow and there are no individual storm peaks unlike for the River Creef which appears to be more dominated by overland flow or rapid throughflow (e.g. via macropores). The River Nol catchment overlies permeable bedrock, is gently sloping, and has soils that enable good infiltration and little chance of saturation to the surface. Therefore, infiltration-excess or saturation-excess overland flows are a rare occurrence here. For the River Creef, however, the soils are thin and sit above impermeable bedrock and so there is frequent saturation of the soils and generation of saturation-excess overland flow.

The proportion of rainfall that reaches a river channel may vary from close to 100 per cent to 0 per cent. Seasonal variability in river flow is known as the **river regime** for which there are four major global types. Where snow and ice melt dominate then there can be a major peak of river flow during the late spring in the case of snow melt or early summer with annual glacier melt. River discharge (the volume of water passing a point in the river per unit of time) can be extremely low during the winter months even though precipitation may be continuing as this precipitation is stored on the glacier in winter. There can also be a strong daily change in river discharge due to daily melt cycles of the snow and ice. Night-time discharge tends to be much lower than that of the mid-afternoon.

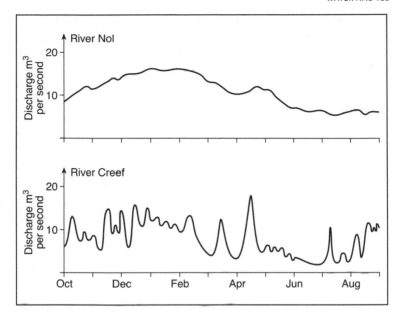

Figure 4.1 The changes in river discharge over a year for two nearby rivers with similar rainfall but with different catchment characteristics.

Arid zones, especially in subtropical drylands, tend to experience very occasional but intense rainfall events. Intense rainfall along with little vegetation cover produces infiltration-excess overland flow, rapid runoff and high flood peaks. However, many dryland soils are coarse and sandy with high infiltration capacities resulting in little chance of overland flow. Therefore, there is a wide variation of response even if rainfall intensities are very high. In most drylands river flow will stop within a few days of the rainstorm and water often seeps into the river beds or is evaporated.

In temperate, oceanic areas precipitation occurs all year perhaps with seasonal maximums (see Chapter 1). The river flow regime in these areas can change in response to either seasonal changes in groundwater storage and release, or higher evaporation and transpiration rates during the summer months.

Rivers in equatorial areas tend to have a fairly regular regime while tropical river systems outside of the equatorial areas receive high precipitation during the summer but experience a marked dry season during the winter. Evaporation and transpiration is high at all times so that the streamflow mirrors the seasonal pattern of rainfall.

Land management change can dramatically alter the flows in river systems. Building dams and changing or diverting stream courses alters river regimes. Deforestation, or intense grazing may result in a large reduction of the infiltration capacity of a soil and decrease in transpiration rates. Covering more of the landscape with concrete and tarmac, which are impermeable, and then channelling flow into drains which feed streams may lead to increased flood risk. These processes lead to more water flowing to rivers with shorter lag times and therefore potentially higher flood peaks (Box 4.2).

Box 4.2 Floods

Flooding is a natural phenomenon and should be expected. Each year flooding causes hundreds of deaths around the world and significant impacts on infrastructure. However, it is staggering when there are flood events such as that in 1998 on the Yangtze River in China which left 14 million people homeless. Floods commonly occur when rivers overtop their banks. However, flooding can also occur through a very high tide at the coast, often made worse by a storm event and high river flows. Flooding may also happen when very heavy rainfall cannot escape from an area (e.g. an urban area where the drainage system cannot cope with the excess surface water) or when there is a lot of saturation-excess overland flow. Flooding can also bring benefits to farmland through supplying nutrients. There is always a flood occurring somewhere in the world (see http://floodobservatory.colorado.edu for maps of current floods).

Often our response to flooding is to build larger and better flood defences around our towns and cities. However, the flood water has to go somewhere and sending the water more quickly through one part of the river system by building large embankments or straightening and deepening river channels simply reduces the lag time downstream and increases the overall flood peak there. Working floodplains mitigate against worse flooding downstream by acting as a temporary store of water. However, historic and continued building or farming on the flat fertile floodplains of the world means that less floodplain can be utilised by a river for this purpose. Flooding in lowland areas is therefore a bigger problem as the extra water is brought downstream more quickly and in greater quantities than ever before. Solutions therefore need to be mixed and include land management solutions as well as flood defence solutions. Dealing with the flood problem means that we need to build resilience into the built environment in flood-prone areas. An example of resilience might be designing buildings that have very flexible ground floor use such as parking areas for cars. When there is a flood warning then the parking area can be evacuated of cars and flooding allowed to occur without any damage to residential dwellings on higher levels.

Flood risk is often interpreted by examining the historic frequency of flooding. If a water level greater than 10 metres in height occurred five times in the past ten years we would say that the return frequency of the 10 metre flood is on average once every two years. However, this does not mean that a 10 metre high flood will occur once every two years. It may be that in one year a 10 metre sized flood occurs three times. Looking at past records might not be the best guide to predicting future flooding, particularly if land management change has taken place in the catchment or if climate change is likely to change the precipitation patterns or vegetation cover.

River channel change

Rivers often have channels that look very different from one another. The shape of river channels also varies within the same river along its course. Most rivers have a long profile (slope of a river from its source to mouth) which is concave with progressively lower gradients downstream. The long profile varies with geology, tectonics and variability in runoff. Other profile shapes also occur when there is interruption by lakes or very resistant rocks (which often result in waterfalls) or through large changes in sea level. If sea levels fall then the whole river may start to erode its bed downward in response. If sea levels rise the river may deposit more sediment and bed levels will be raised along its length as downstream processes have knock-on effects upstream, as well as upstream processes delivering water and sediment downstream.

Rivers are dynamic in that they often move position around the landscape through time, change their shape and move sediment, water and dissolved materials. They are an important agent in removing weathered material from the land surface and redistributing it across the landscape or into oceans. This process balances out the mountain building caused by plate tectonics so that over long time periods plate tectonics might build mountains but weathering, erosion and removal by river systems of ice masses (see later in this chapter) smoothes out the landscape again. Uplift and erosion of landscapes has traditionally been thought of as cyclical, although the outcomes in each environment may be very different depending on rates and types of processes operating there.

Sediment movement in rivers is linked with water flow. If a piece of sediment is to be picked up from the river bed or bank by flowing water a critical threshold has to be passed above which the water velocity or **shear stress** is sufficiently large to overcome frictional forces that resist erosion. The transport of materials close to the bed is known as bedload transport. Particles move by rolling, sliding or **saltating** (hopping) along or close to the bed. If the flow velocity does not change, a particle will only come to rest if it becomes lodged against an obstruction or falls into an area sheltered from the main force of the water by a larger particle. With further increases in the strength of flow, the smaller particles may be carried upwards into the main body of water and transported in suspension. Deposition and cessation of movement for an individual particle occurs when velocity falls below critical conditions. This means that finer particles are preferentially moved downstream. For suspended sediment within the water body (as opposed to that transported close to the river bed as bedload transport), transport is determined not only by water flow but also by the rate at which it is supplied to the river (e.g. from washflow process).

Where erosion exceeds deposition within a particular section of a river, there will be lowering of the river bed or widening of the banks. If erosion and deposition occur at the same rate then the river will stay at the same level. Eroding channels may undermine structures such as bridges while depositing channels may submerge structures such as roads. Stable channels, especially those whose beds are lined with bedrock, are less likely to be a problem to engineering structures but fluctuations in river channel dimensions and locations caused by flooding or sediment pulses moving down the river can be problematic.

River forms are often described by features of the cross-section of the channel as if it were full to the top with water. Often channel cross-sections adjust to accommodate discharge and sediment loads. Channels are expected to be wider and deeper if the discharge is greater. As you move downstream, river channel cross-sections tend to get larger although there is not a perfect relationship between discharge and channel cross-sectional area because larger channels are more efficient at carrying water (less friction around the channel edges per volume of water). Furthermore, the sediment that makes up the river channel is important. Channels with a high percentage of silt or clay in their banks (which are often more characteristic of lowland sections of river), and rivers transporting much of the sediment load in suspension, tend to be narrower and deeper than sand and gravel-bed rivers. Vegetation can also be important in controlling cross-section shapes by influencing bank resistance through root systems that bind the sediment. Removing bankside vegetation can lead to rapid bank erosion.

Within channel cross-sections there can be large variations in the velocity of water flow. Close to the bed or banks of the river the velocities tend to be slower. On bends of rivers there can be forces exerted that increase the pressure on the outer bank as the water flows by with less pressure on the inner bank of the bend. Bank erosion is more likely on the outer bank as faster water has more chance of picking up sediment. On the inner bank, sediment is more likely to be deposited. This causes further development of the **meander** (Figure 4.2) and means that the river is continuously eroding and depositing sediment as a natural process and so the exact position of the river bank will change through time. Within meandering channels, because water velocity is greatest close to the outer bank, sediment size tends to be greatest here and slowly decreases towards the inner bank. When meander bends become too exaggerated the river will cut through the meander to a direct course downstream for that short section, leaving behind a crescent-shaped section of channel that is cut off from the main river, known as an **oxbow lake**.

Looking at a river directly from above in a plane you can see the pattern, or **planform**, of the river. River patterns occur mainly

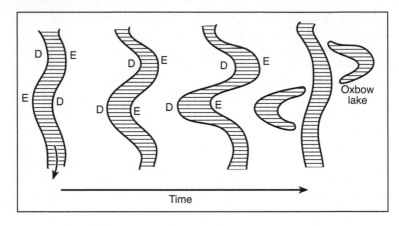

Figure 4.2 The enlargement of river meanders until the river short cuts the meanders creating an oxbow lake. E is a zone of net erosion and D is a zone of net sediment deposition.

as braided, meandering and straight channels and a single river may have each of these patterns in different sections. Braided channels consist of lots of individual smaller channels which separate and come together with the islands in between known as bars. The channels can rapidly move locations with bars eroding on one side and growing on the other. Braided rivers are common where there is a lot of mobile sediment such as downstream of a glacier. Meandering channels have a wavy planform. Often waviness is measured by sinuosity to determine the nature of the measuring channel where sinuosity is the ratio of the length of river between two points compared to the length of the straight line valley between these two points. Straight channels are defined as having a sinuosity of less than 1.5. A meandering channel refers to a single channel with a number of bends which result in a channel sinuosity of 1.5 or greater. Straight rivers are more controlled by human action and natural straight rivers are often unstable and become meandering. Even a drop of water running down the surface of a perfectly smooth piece of glass will take a meandering course.

Channel slope and discharge are important controlling variables on the planform of river channels. For a given discharge there is often a critical slope above which channels will meander and then a further threshold above which they will braid. These thresholds decrease with increasing discharge. Thus, braided sections are usually found on large rivers or on small rivers with steep slopes. Wandering and braided rivers will occur where there is coarse sediment, erodible banks and where the main sediment transport mechanism is bedload transport rather than suspended transport.

The characteristics of river beds tend to change downstream. Often there are bedrock channels in the upper section of a river network or large boulders and cobbles. There is a sharp decline of bed material size downstream. This pattern is because smaller particles are transported downstream more easily and because of abrasion of larger sediments within the river by colliding and grinding causes material to get smaller and more rounded downstream. However, these patterns are not seen everywhere and may be disturbed by local sediment inputs to the river network.

Within rivers there are several erosional and depositional landforms. In bedrock channels potholes can be found formed by: mechanical wearing and grinding of small particles enlarging an existing small depression or weakness in the rock; pressure changes due to bubble collapse in turbulent flow; and chemical weathering. Within rivers that have gravels on their bed the most common landforms on the bed are sequences of pools and **riffles**. Pools are deep sections with relatively slow flowing water with fine bed material. Riffles are formed by accumulation of coarse sediment with shallow, fast flowing water. The spacing of pools and riffles is often five to seven times the channel width but these do vary. The bed of sandy sections of rivers can have small ripples in the sand which are less than 4 centimetres in height and then also larger dune features. The size and shape of these dune features change with discharge during rainfall events. The dunes and ripples tend to migrate downstream as sand is carried up the upstream facing side of the ridge of the feature and then falls down the downstream facing side. At very high flow velocities, a flat river bed can be formed or dunes can even migrate upstream since erosion from the downstream side of the dune allows suspension of material in the water which occurs faster than it can be replenished from upstream.

River channels change their slope, cross-section, planform, and bed forms in response to environmental change. Humans have modified river channels over the past few thousand years although changes during the last two centuries have been greatest. Activities such as dam construction, urbanisation, mining, land drainage and deforestation can all impact river channels. Faster, more peaky flow from urbanisation or deforestation can accelerate erosion. In the USA, channel volumes have been found at up to six times the size of those similar rivers with more natural flows. Dams are having a major impact on river flows, channels and sediment dynamics. For example, the Nile now only transports 8 per cent of its natural load of silt below the Aswan Dam thereby reducing the fertility of the downstream flood plains and accelerating river bed and coastal erosion as the lack of sediment entering the sea no longer replenishes the sediment being eroded by wave action (see section on coastal processes below).

The above discussion has shown that river channels are not static features in the landscape. Despite this, rivers often act as boundaries for land ownership which can lead to disputes because the river course changes through time and so the boundary between property, counties, states and countries can change too. Understanding the processes and dynamism involved in river channels is essential to good management otherwise great expense might be incurred when engineering structures fail due to river channel change. There are many examples over the last century of engineering failures around rivers. A notable example of this is the Mississippi River, which regained much of its sinuosity following engineered channel straightening in the early twentieth century. Indeed, many rivers that have been subject to engineering features, such as straightening, became quite sterile environments as the biological variety within the river was removed so that wildlife suffered. Many of these rivers are now being rehabilitated to try to restore features that encourage greater wildlife (such as meander bends with pools and riffles) and also to work with the erosion and deposition processes within rivers rather than against them.

Water quality and pollution

Water quality is a measure of the various chemicals that are found within the water. Water quality can vary naturally and be influenced by human action. Water pollution occurs when the water environment is changed so that the species using the water can no longer tolerate its chemistry and either die or quickly leave that section of the water body. Often, water quality and pollution are measured from a human perspective in terms of how safe the water is to drink or how easy or costly it is to treat the water before it can be consumed. Water that has a good taste and is not dangerous to health still contains dissolved substances. The highest quality waters with the best taste tend to come from reservoirs and lakes that have collected the majority of runoff from undisturbed landscapes with mainly surface water runoff. However, many groundwaters can taste good too. These tend to be where rocks only weather slowly. If groundwater has picked up lots of dissolved substances from the surface, soils and rocks it is more likely to have an unpleasant taste (see Box 4.3).

The runoff routes for water across and through the landscape are important for controlling the concentrations of different chemicals within rivers, lakes and groundwater. As precipitation inputs normally have low concentrations of dissolved chemicals where a site is dominated by rapid water movement to rivers (macropore flow or infiltration-excess overland flow) there is little time for chemical interactions with soil or rock and so river water chemistry will be similar to rainwater chemistry (e.g. in blanket peatlands). However, if there is significant overland flow which brings lots of sediment with it, perhaps with other added fertilisers or industrial chemicals, then that can cause pollution problems. Throughflow in upper soil layers and water coming back out of the soil as part of saturation-excess overland flow will tend to have quite different solute concentrations to that of precipitation since it will have had more time to interact with the soil and for weathering reactions to take place. Solute concentrations in groundwater are often greater than elsewhere due to the longer contact time of water with soil and rock. The groundwater composition is affected by the geochemistry of the surrounding geology since different minerals weather at different rates and produce different solutes.

Water quality varies with time because the water flow processes across the landscape also vary with time. As groundwater and throughflow deep within the soil are the major source of base cations to river water from mineral weathering, base cation concentrations in river water tend to be greater during dry conditions when groundwater is the main source of river flow. In wetter conditions more flow is generated from macropores and over the land surface containing smaller base cation concentrations. The availability of solute supply can vary with temperature during the year or with plant growth (e.g. nitrate concentrations can be lowest in spring and summer due to plant uptake). The concentrations of chemicals within precipitation can also vary with time. In coastal regions, precipitation is often enriched by sea salt during stormy weather, increasing chloride concentrations in river waters.

Solute concentrations and **solute fluxes** (total mass of a solute moved) in rivers and lakes vary locally and globally. This is due to local and regional differences in climate, geology, topography, land management, soils and vegetation. Globally, patterns of solute concentrations are dominated by climate and geology. On a regional level, land management, soil types and vegetation are more important. Land use affects solute concentrations by altering runoff pathways and the amount and availability of dissolved chemical sources. It has been estimated that human action has increased the total amount of solutes transported by rivers across the planet by 12 per cent.

Agriculture has had a large impact on water quality, particularly through increased erosion, and the leaching of nutrients, pesticides and by-products of veterinary medicines into water courses. There may be individual point sources of chemicals from leakage of pesticides, slurry and wastes from storage facilities, or from more diffuse pollution across the landscape. Manure is often spread on farmland but when the plants cannot take up all of the nutrients provided (i.e. too much has been applied) or if the manure is applied just before heavy rain then leaching of soluble nitrogen and other chemicals occurs. Drainage and ploughing can also increase leaching rates and erosion.

Urbanisation and industrialisation are associated with increased river concentrations of metals, nutrients, organic matter and salt in the winter in cold climates where roads are treated. Vehicles are responsible for a lot of urban metal pollution as they corrode depositing materials on the urban surface. Urban drainage with rapid removal of surface waters to rivers also means that there can be a strong flush of chemicals that have built up on the surface at the start of a rainfall event. Sewage effluent is also another major pollutant. In many developed countries there are sophisticated wastewater treatment techniques that clean the water before it is returned to

Box 4.3 The groundwater quality time bomb

There have been many instances where groundwater contamination has been related to damage to human health (e.g. too much fluoride causing debilitating bone diseases, or too much nitrate causing lack of oxygen in the blood; arsenic poisoning which seems to be an acute problem in Bangladesh and West Bengal, India). Often it is long-term exposure that is the problem rather than a one-off drink. This is because groundwaters, unlike surface waters, are often fairly clean from sources of infectious disease. However, this is not always the case, especially if there is poor waste water management on the surface. It also depends on how good a filter the soil and rock are within the aquifer. Indeed this filtering property means that supplies from local wells are often not chlorinated to remove the infectious carriers such as *E. coli* bacteria.

Human health problems have led to legal standards for the concentrations of many solutes in drinking water within most countries. Contamination can take place quite some distance from the abstraction point if the aquifers are well connected. Surface landfills, chemical spills, leaking underground tanks at fuel stations and many other surface management strategies lead to contamination of groundwater from point sources. Diffuse sources such as fertilisers and pesticides from farming may also be important contaminants in some areas. In many places there is only a slow percolation of contaminants into some aquifers. For some deep aquifers the water may take 50 years or more to percolate down. In some developed countries the peak use of fertilisers that were applied at rates that were too great for plant uptake occurred in the 1970s or 1980s. Here, despite good environmental protection measures since then, and careful controls over the use of fertilisers in groundwater source zones, the levels of solutes such as nitrate may be continuing to increase in groundwater. There is therefore a long time lag between surface management and groundwater quality changes in some locations so that the groundwater problem only emerges several decades after good management has been implemented. One solution to this is to mix the contaminated groundwater with water that has relatively low concentrations of solutes to enable all of the water resource to be used. However, this would require multiple water sources for a supply area, which is not always possible. Instead, very costly removal techniques have to be applied to strip out the contaminants from the water.

rivers. However, during major storm events many of the sewer systems cannot cope with the amount of water entering them and so pollution of river waters occurs as the sewage systems overflow. Even with sophisticated treatment this may not remove a large range of compounds that have entered the water system through human use such as medicines, antibacterial products, disinfectants, antibiotics, fire retardants and some chemicals in soaps, shampoos and other personal care products. Some of these chemicals may be toxic to aquatic organisms. Other chemicals others such as hormones from certain drugs may influence other aquatic processes including reducing the proportion of male fish compared to female fish.

In an attempt to tackle water quality problems in holistic way the European Union devised a legal process called the EU Water Framework Directive. This is described in Box 4.4.

Box 4.4 The EU Water Framework Directive

The Water Framework Directive is a legally binding process for assessing and improving the quality of surface waters in rivers and lakes, groundwater and coastal waters. These water bodies have to be seen to be achieving good status and, if they are not, efforts have to be demonstrated to improve the status of the water bodies through whole river basin management plans. European Union countries that fail this process may be fined. The key innovation is that whole land management solutions are expected to deal with diffuse and point source pollution and the provision of sustainable amounts of water within rivers and groundwater. Assessments have to include not just measures of water quality but additional measures such as the status of river beds, the ecology and the flow regime of the river, the sustainability of the groundwater resource and so on, in comparison to ideal good examples of what the system should look like under natural conditions. In reality this is all very challenging to achieve, not least because every water body is unique due to local soils, vegetation, slope, geology and climate. Therefore, finding reliable pristine examples which other water bodies must emulate is not straightforward. However, the general principle of having an integrated approach to water management is an important one whereby land owners, regulatory agencies and government bodies can work together to come up with solutions across the landscape, rather than with dealing with just the symptoms of the problems within the water body itself.

COASTS

Coastal areas are also subject to water pollution, much of which is derived from polluted river waters entering the coastal environment as well as pollution from ships, oil spills and some industrial effluent that is discharged into the sea. Coastal environments are important since 40 per cent of the world's population live within 100 kilometres of the coast and it is even thought that 75 per cent of the world's population could be living within 60 kilometres of the coast by 2025. Coastal environments support fisheries, large ecosystems, leisure activity and even power generation and act as an important buffer during storm events. As discussed in Chapter 2, sea level rise is a major threat to the coastal community. Sea levels are currently rising at 3 millimetres per year with a predicted rise of 18 to 59 centimetres by 2100 leading to increased flooding and erosion of land. It is therefore important to understand the physical geography of coastal regions so that we may better anticipate and adapt to changes in the future.

Waves

Waves are the most important feature of coastal environments as they drive many sediment transport processes and therefore the inputs and outputs of sediment from an area. The inputs and outputs of sediment in turn help shape the landforms of the coastal environment. Waves are generated by wind. Stronger wind results in larger waves. Waves can travel vast distances across the oceans and so wind conditions a considerable distance from the coast can influence coastal waves. Waves can be measured for their height H (height between top and bottom of the wave surface), length L (distance between successive wave peaks (crests)) and period (i.e. time between two wave peaks or troughs passing by a point).

Wave behaviour depends on the depth of water in the ocean. Where the water depth is more than twice the wave length then as waves travel across the water surface, the water beneath moves in circles; forward under the crest of the wave and backward under the trough. The diameter of the circles of water decreases with depth until in deep water the wave motion is no longer detectable. In water which has a more intermediate depth, where the depth is between twice the wave length and a twentieth of the wavelength, then the wave motion will be affected by friction on the seabed. This means that the water motion beneath the waves becomes more elliptical with the ellipses being smaller and flatter nearer the bed so that at the bed water just moves back and forth. In shallow water very close to the shore where the water depth is less than a twentieth of the wavelength then the water movement is just horizontal in a back and forth direction. In reality what this all means is that as waves move from deep to shallow water they get closer together and slow down. Waves also get higher as they travel into shallower water, a process known as **shoaling**. They also change their shape from a more symmetrical wave shape to a shape with more peaked crests and flatter troughs. In deep areas, water movement produced by waves is back and forth at the same velocity (i.e. it is the shape of waves and their energy that is moving across the ocean but not the water itself within the waves). However, close to the shore the onshore side of the wave is stronger and this promotes sediment transport more favourably in the onshore direction. Wave **refraction** occurs as the wave moves close to the shore. This is when the section of the wave in shallower water travels slower due to friction on the bed than the section of the wave in deeper water. The outcome is that the wave crest rotates to become parallel with the contours of the seabed so that wave direction 'bends' as it approaches the shore.

When the water depth becomes too shallow for the wave then the wave breaks up. When the water depth is just slightly greater than the wave height then the wave breaks up creating an area known as the **surf zone**. The energy released by breaking waves can be significant and help generate nearshore currents and sediment transport. When waves break they cause a rise in the water level on the beach with the returning water running back down the slope into the sea. This motion is called swash and the up-beach movement is better at transporting sediment than the returning flow and so this helps maintain the gradient of the beach.

Nearshore currents are important for landform development in coastal areas. These currents gain their energy from wave breaking so if the waves are stronger (e.g. during storms) then the currents will be stronger. Longshore currents flow parallel to the shore within the surf zone driven by waves entering the surf zone with their crests aligned at oblique angles to the shoreline (Figure 4.3). They are also affected by winds and longshore currents can be particularly strong when winds are blowing in the same direction as the longshore current. Longshore currents can transport large quantities of sediment, often known as **longshore drift**. The currents can be fast flowing and this is assisted by sediment being stirred up by wave breaking, making the sediment easier to pick up and transport. Another type of nearshore current which all those who go swimming or surfing in near shore coastal waters should be aware of is the rip current. These are strong, narrow, seaward flowing currents which flow back through gaps between sandbars. The water piles up as a result of the waves approaching at an oblique angle and this water rushes back through the surf zone at key points.

A final important wave type is that of a tsunami. These can result in an enormous wave which floods coastal areas and causes massive loss of life such as in the event on 26 December 2004 around coastlines of the Indian Ocean which killed an estimated quarter of a million people. That particular tsunami was caused by an earthquake which displaced water deep within the ocean. Tsunamis can also be caused by a large landslide entering the sea such as that which occurred after the eruption of the Krakatoa volcano

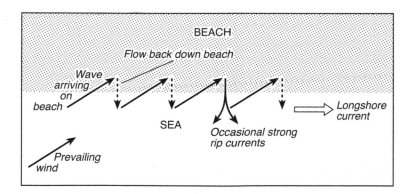

Figure 4.3 Longshore current produced by the combination of waves arriving at oblique angles to the sea perpendicular to the shore. The result is an overall movement of water and sediment along the coastline.

in 1883 killing 33,000 people. When a tsunami travels across the deep ocean it can typically only have a height of a few centimetres (60 centimetres after two hours on 26 December 2004) but they travel quickly at around 500–1,000 kilometres per hour. When they reach shallower water and approach the land a tsunami wave will begin to shoal, causing the water to suddenly retreat a long way from the beach and eventually when the wave arrives it can reach tens of metres in height.

Tides

Tides are driven by the gravitational pull of the Moon and the Sun on the Earth. Tides have a predictable daily and monthly cycle. The Earth and Moon exert a gravitational pull on each other, which is counterbalanced by forces associated with the Earth's orbital rotation. In theory, the force exerted due to the Earth's rotation is equal on all parts of the Earth's surface except where the gravitational pull of the Moon allows for a slight reduction. This reduction allows the oceanic surface to bulge. This bulge will occur on that part of the Earth closest to the Moon. In addition there is a balancing bulge on the opposite side of the Earth as a product of the forces exerted by the rotation of the Earth. Consequently there are two bulges and thus two tides per day. The tide rises and falls at a point on the Earth's surface as that point rotates away from and towards the direct line of the gravitational pull of the Moon. The gravitational pull of the Sun introduces an extra monthly dimension to the tidal sequence. As the Moon's position relative to the Sun changes over the lunar month (the Moon takes 28 days to revolve around the Earth) the two astronomical bodies either pull in alignment or opposition to each other. During the full and new moon phases they pull along the same direction allowing the oceanic surface to bulge further resulting in larger magnitude tides known as **spring tides**. During the half moon phases the Sun and Moon pull in opposite directions resulting in subdued tidal bulges and these are known as **neap tides**. The actual impact of the gravitational pull on tides depends on the shape and topography of the coast. In some locations the tide can rise and fall by several metres whereas in others there is a barely noticeable tide. The largest tidal ranges seem to be associated with constricted areas where there are narrow seaways connecting the sea to the main ocean such as the Bay of Fundy, Canada (17 metre tidal range) or the Irish Sea (13 metre tidal range in the Severn Estuary) or where there are wide continental shelves, such as off the east coast of China. On coasts facing the wide, open ocean, tidal ranges are usually less than 2 metres.

As the tide rises and falls against the coastline it produces a flow of water. This is a **tidal current**. If you throw an orange into the water at the shoreline during a rising tide the orange is pushed up the beach. However, if you throw it in at the same point on a falling tide the orange will float out to sea as the tidal current takes it away from the shore and you will lose your orange. The geometry of the coastline can control the tidal current which is most pronounced in river mouths, estuaries and where flow is squeezed through inlets.

Coastal landforms

There are a number of driving forces behind coastal landforms. These can be grouped into waves, tides and rivers, although obviously the nature of the land material at the coast is also important (e.g. hard rock or erodible loose sediment). There are different landforms that are characteristic of locations where waves, tides or riverine processes dominate. For example, a common feature of wave-dominated coasts is a beach perhaps with coastal dunes.

Beaches are formed by deposits of sediment brought by waves and are typically shaped with a concave profile. Towards the top of the beach is the **berm** which is where the slope steepens and then flattens off. Within the beach there can often be cusps of sand or gravel at the shoreline which are repetitive features a few metres apart formed by swash action. Beaches respond to changing wave energy conditions. When it is calm and wave energy is low then the net sediment transport is in an onshore direction resulting in steepening of the beach and a pronounced berm. In stormy conditions net offshore sediment transport occurs with the destruction of the berm and flattening of the beach. This then helps to dissipate the wave energy over a wider area.

Coastal dunes protect the coastal area behind them by providing a buffer to extreme waves and winds. Dune formation requires wind and a large supply of sand. Onshore winds (see Chapter 1) capable of sediment transport must occur for a significant amount of time. Dunes often develop just above the spring high tide line where litter such as seaweed and wood collect. This litter starts to trap sand blowing around and promotes small dunes to form. Once these start then they can enlarge, especially when plants grow within them allowing a greater height of sand to be trapped around the plants. Dunes can grow relatively quickly under the right conditions reaching 2 metres high after just five years.

Just off the coast in wave-dominated environments there can be formation of barrier landforms including barrier islands and lagoons. These extremely dynamic landforms can be found along about an eighth of the world's coastline and there are many famous examples such as those along the Florida Atlantic coast or the coast of The Netherlands. Barriers help buffer inland areas from storm wave energy and represent a large accumulation of onshore moving sand bars. They form long strings of parallel island chains punctuated by tidal inlets that allow the transfer of water and sediments between the open sea and the lagoons behind the barrier. Some barriers are aligned to the swash direction whereas others are aligned to the prevailing direction of long shore currents. For example, a **spit** is a narrow accumulation of sand or gravel, with one end attached to the mainland and the other projecting into the sea or across the mouth of an estuary or a bay. Spits grow in the long shore drift direction and can only exist where there is a regular supply of sediment.

Estuaries are river mouth locations where sedimentary deposits from both river and sea sources create landforms. Sea levels rose around river mouths as ice melted following the last glacial period and this stabilised around 6,000 years ago. Infilling of estuaries occurred as sediments from land and sea could inundate the new deep waters at the river mouth. Estuaries have three broad zones; an upper, middle and outer part. River processes dominate the upper estuary while marine processes dominate at the outer estuary. The middle section is mixed. The same volume of sea water leaves the estuary during the falling tide as enters during the rising tide. However, the duration and strength of the rising and falling tide tends to be different. Estuaries or channel sections that display a flooding tide that is faster and stronger than the ebbing tide are said to be flood-dominant, whereas those that display an ebbing tide that is greatest are ebb-dominant. Flood- or ebb-dominance influences whether net sediment transport is landward or seaward, respectively. Flood-dominant estuaries infill their entrance channels by continually pushing coastal sediment landward, often causing the exit to be clogged up, and therefore the estuary mouth changes which makes their use for shipping difficult. Ebb-dominant estuaries tend to flush sediment seawards and are more stable environments for shipping.

The outer zone in most estuaries is devoid of vegetation due to excessive sediment and water movement. Further within the estuary, however, salt tolerant plants can grow. In tropical environments this leads to mangroves and in temperate environments to salt marshes. These significantly enhance sediment deposition. The steady supply of organic matter from the plants also adds to the sediment, which means that these zones rise in altitude over time as the sediment accumulates. This build-up of sediment is often at a faster rate than sea level rise meaning that these environments can rise at the same rate as sea levels, as long as they are protected from human damage.

Deltas are coastal landforms dominated by river processes. They are accumulations of sediment deposited where rivers enter into the sea. Here the amount of sediment delivered by the river is greater than that removed by waves and tides and so the delta moves seawards. Lots of people live on deltas, although there are not many deltas around the world as they tend only to be associated with large river systems (e.g. Mississippi, Ganges, Lena, Nile). The coarsest sediments are deposited close to the river mouth and the finer sediments settle out further seaward. Ongoing delta survival and development relies on an active sediment supply by the river. Deltas can be very dynamic. As one area of the delta grows upwards due to building sediment, the river channels can suddenly no longer keep flowing to that raised area and so they move across the delta supplying sediment elsewhere. The starving of sediment from some areas of delta can then lead to net erosion through wave and tide action. This dynamism creates a hazard for humans living on deltas. Humans can also modify the sediment supply in a river, by building dams for example, which can lead to delta erosion.

On the other side of the spectrum, rocky coasts may be seen to be solid, more stable landforms. However, they are actually characterised by erosional features. This often makes for stunning scenery with rugged cliffs and interesting **stack** features, arches and caves. Nevertheless, the average rate of change along rocky coasts is slow, although some changes such as landslips can be dramatic. Rocky coast erosion occurs through mass movements, rock weathering processes and rock transport processes. Mass movements are common due to steep slopes. Rockfalls are characteristic of hard rocks, landslides typically occur in thick deposits of clay, shale or **marl**, and flows occur when there is a high liquid content. Freeze-thaw, wetting and drying, chemical weathering and the mechanical abrasion and force of water undercutting the base of rocky slopes by wave processes are principal drivers of mass movement. These processes were discussed in more detail in Chapter 3. Typically, once material has been removed from the cliff face onto the floor below it can be removed by coastal transport processes, meaning that the bases of cliffs do not receive protection for long by debris deposited through mass movements. **Shore platforms** develop when erosion of a rocky coast leaves behind a horizontal or gently sloping rock surface.

Coral **reefs** are depositional environments yet are located in high energy wave systems. They are the largest biologically constructed formations on Earth and consist of limestone that has been created by animals forming their shells. When corals die, they leave behind the limestone that formed their skeletons and so the sediment can build up over thousands of years. Coral reefs are in a very delicate balance between erosion and biological construction. Without the presence of living organisms the reefs would not exist because weathering and wave erosion would destroy the landforms. Corals can be found throughout the world but reef building corals are only found in the sub-tropics between 30°N and 30°S.

Coral reefs occur in two main settings. The first is close to land on the continental shelf where water depths are less than 200 metres such as in the Great Barrier Reef off the north-east coast of Queensland, Australia. The second are those that rise several kilometres from the ocean floor. These have formed around the edges of volcanic islands above hot spots (see Chapter 3 for an explanation of oceanic island formation). Often when the volcano becomes extinct, the island erodes and also sinks back into the ocean because the oceanic crust on which the volcano rests will cool and sink over millions of years as the crust moves away from the seafloor spreading centre at the mid-ocean ridge. However, coral growth may be able to keep up with the rate of island sinking and so there may be a large depth of limestone formed on top of the volcanic base. The key is whether vertical reef growth can keep pace with falling land levels or rising sea levels. **Atolls** are reefs that surround a central lagoon and most of these are found in the Indian and Pacific Oceans. They tend to be circular in shape and range from 75 kilometres width to less than 1 kilometre. Many are believed to have developed around the rim of a volcanic island. As the volcano has itself sunk back into the ocean, the coral has been able to keep up growth but only around the edges where the coral had previously developed. Thus the central portion of the island is largely devoid of reef and a lagoon forms. The nutrients in the lagoon are poor as wave action is restricted and thus growth is limited.

Coastal management

In addition to sea level change coastal managers have to deal with natural erosion and depositional processes and those processes that have been created by human action (e.g. delta erosion due to upstream dam construction; beach erosion due to sand abstraction somewhere further down the coast which starves longshore drift of sediment). Many coastal management solutions have involved engineering solutions such as sea walls, breakwaters and **groynes**. Sea walls are large and costly concrete, steel or timber structures often with a curved face. However, as sea walls provide a limit to the beach zone they may stop the usual process of beach profiles changing (lengthening/flattening) during stormy periods and steepening during calm periods. Sea walls also reflect waves and instead of dissipating energy more erosion takes place at points further along the coastline beyond the extent of the wall. A common response is to then extend the seawall further but this just shifts the erosion problem further along the coast. A sea wall was built in the nineteenth century at the seaside resort of Blackpool, England and 150 years later almost 80 kilometres of coastline in the area had a sea wall as erosion developed along the coast from the Blackpool wall.

An alternative strategy to sea walls is to reduce the incoming wave energy by installing submerged breakwaters parallel to the shore that encourage waves to break further away from the land. These features must be porous so that they allow sediment to move through them and are usually built as a series to protect a stretch of coast. They are very expensive to install because they must withstand extreme wave action and are placed within the most energetic part of the nearshore zone. Building beaches is another solution to coastal erosion. This involves artificial deposition of sediment on the beach or in the nearshore zone in order to advance the shoreline seaward. However, this sort of approach can often treat the symptoms and not the causes of the erosion problem and beach nourishment can simply be followed by net erosion. Officials in Miami Beach, Florida spend millions of dollars every few years transporting sand to replenish the beach. Generally the sand used for beach nourishment must be coarser than local sediment in order to minimise rapid sediment loss offshore. Groynes are often installed in beaches to trap sediment moving via longshore drift. The groynes can be buried by sediment allowing longshore drift to recommence. A problem with groynes is that rip currents sometimes develop on their downdrift side which consequently moves sediment offshore and away from the beach system. Therefore some groynes constructed with bends to counteract this effect. Jetties are built to line the banks of tidal inlets or river outlets to stabilise the waterway for navigation. However, these jetties often encourage deposition on the updrift side and erosion downdrift. This occurred in the 1920s at Santa Barbara, USA, where the harbour infilled with sediment. Further down the coast, as a result, the whole community became at risk from coastal erosion.

Overall there are probably four main strategies for dealing with coastal erosion and sea level rise. These are: doing nothing, abandonment, adaptation or protection. The first option might be most costly as it leaves infrastructure and people at risk and so can only be realistic if the area is sparsely populated. Abandonment is where people and industry leave the coastal setting and move inland and also where further coastal development is prevented. Adaptation involves designing features into the landscape to cope with change such as building homes on stilts or providing warning systems.

Finally, protection involves engineering solutions such as tidal barriers with large sluice gates on estuaries to prevent very high tides from entering the estuary. The challenges for coastal management are enormous because the system is dynamic requiring flows of sediment, water and energy. Interfering with those flows in one place has knock-on effects further along the coast. Some coastal areas are currently responding most to short-term changes caused by human action whereas others are still responding to changes in sea level since the retreat of ice sheets began at the end of the last glacial period some 18,000 years ago.

ICE

Ice exists on Earth today mainly in large ice sheets over the Antarctic and Greenland, in ice caps, in sea-ice over the North Pole and in valley glaciers on land. It also exists in large quantities within frozen ground known as **permafrost** and produces characteristic landforms in cold regions. If the amount of ice present today in the Greenland ice sheet melted this would raise sea levels around the world by 5 metres. Ice sheets form over very large areas, the size of continents, and are typically a few kilometres thick. Ice sheets flow slowly although there can be faster moving rivers of ice within them. In Antarctica the faster moving ice streams feed **ice shelves** which are formations of floating ice which then melt into the ocean or are broken away and float off into the ocean.

Glacier and ice sheet dynamics

Glaciers are much smaller than ice sheets, covering individual valleys on land but they occur on every continent. Glaciers are formed in mountainous regions wherever snow accumulates at a faster rate than it can be melted. This therefore requires both a good supply of precipitation and cold conditions. A typical glacier has an accumulation zone at the top where rates of gain are more than losses and an **ablation zone** at the bottom where rates of loss are greater than rates of gain. The loss of ice at the outlet of the glacier mainly occurs through melt water but sometimes, if the glacier directly flows into the sea, icebergs can be broken off and float away before melting.

As the melting point of ice reduces with increasing pressure there is an important difference to be made between 'warm' ice and 'cold' ice. At a depth of 1 kilometre within an ice sheet the melting point of ice occurs at the cooler temperature of $-0.7°C$, rather than $0°C$. Therefore, the thicker the ice mass the more likely it is to produce water at depth at the **pressure-melting point**. Warm ice is at the pressure-melting point and contains liquid water, whereas cold ice occurs at temperatures below the pressure-melting point and does not contain liquid water. The pressure-melting point concept is important for understanding what might be happening within the ice and at the base of the ice mass, because water can help glaciers move over the bed. Warm ice occurs throughout temperate glaciers except near the surface of the glacier which becomes cold in winter. Cold ice occurs throughout cold glaciers. If a glacier bed is cold flowing water will not occur and so there will be less sliding and deformation of sediments than under a warm ice base. In some places the heat released from the Earth produced by tectonic activity can melt ice at the base of an ice mass. There are several lakes below the Antarctic ice mass, the largest being 4 kilometres below the ice surface, and covering around 14,000 square kilometres. One of the theories behind why these large lakes exist is that they are warmed by tectonic activity occurring below them. However, little is known about these lakes and they are difficult to study because of their location.

Water is produced at the snow surface of many glaciers in summer. Unless it refreezes, meltwater may percolate downwards through any snow and run across the surface of the glacier ice. Water will flow downslope along the ice surface and will emerge as a series of small streams many of which will have created smooth channels on the ice surface. The water may flow to the outlet of the glacier or some of it may flow within the glacier in tunnels. Water that descends from the glacier surface may arrive at the bed at discrete locations flowing in channels. Two types of channels exist beneath glaciers: Nye (N) channels and Röthlisberger (R) channels. N-channels are formed in the bedrock, and R-channels are formed upwards into the ice.

To move ice from the accumulation zone downslope to the ablation zone the glacier must physically flow downslope. There are three mechanisms by which glaciers flow: internal deformation or creep (sometimes called plastic flow), sliding (sometimes called basal slip), and bed deformation. Stress applied under the action of gravity causes the ice to deform and to creep along. While the creep rate is much lower for cold ice than warm ice, creep is a slow process. Once the pressure melting point is reached at the glacier bed, sliding can occur as water reduces the friction. If the bed of a glacier is cold, the sliding is restricted. Glacier bases often contain rock obstacles. Here **regelation** can occur. If ice flows around obstacles at the bed there will be excess pressure upstream of the obstacle and lowered pressure on the downstream side. Increased pressure lowers the ice melting point upstream of the obstacle. The melted water then flows around the obstacle to the low-pressure downstream side where it refreezes because the melting point is higher. This mechanism therefore allows the ice to slide past the obstacle. Regelation is limited by heat conduction, which is better for small obstacles. The regelation process is important as it often causes obstacles to get frozen into the base of the ice and then moved with the ice mass. Where an ice sheet or glacier sits on soft sediments then movement of these sediments can also assist ice movement as the sediments themselves deform.

The rate of movement of glaciers is very variable. Temperate glaciers often flow at tens of metres per year whereas cold glaciers flow at rates of two metres per year or less. Some fast flowing ice streams can flow at several hundred metres per year by fast sliding or on sediment that is deforming. Around 1 per cent of glaciers experience sudden phases of surging and then quiet slow moving periods. It is uncertain why this is the case.

Glacial landforms

Glacial erosion removes enormous volumes of rock and produces amazing landforms. The mass of ice can crush rock where there are weaknesses, producing angular sediments. Crushed bedrock can be taken up into the glacier. The process by which a glacier removes large chunks of rock from its bed is known as plucking. Plucking can occur by ice regelation around the rock or by incorporation into the ice along faults. Glacial abrasion occurs when rocks and particles at the base of the ice slide over the bedrock thereby scratching and wearing it down. The sediment resulting from abrasion is very fine and when suspended in water is known as glacial flour. Meltwater at the bed of a glacier can also cause erosion by mechanical or chemical processes as with a normal river system. Glacial meltwater is good at erosion as it tends to flow quickly and contains lots of abrasive sediment. The rate at which glaciers and ice sheets erode varies depending on temperature, the rate of ice movement and the bedrock. On average across the planet it seems that a rate of 1 metre of erosion for every 1,000 years is a reasonable estimate for ice masses.

Weathering action such as freeze-thaw and undercutting by the glacier results in material falling from rock slopes onto the surface of glaciers where it is transported downstream by the ice mass. If the material falls onto the glacier in the upper accumulation zone, it will become buried within the glacier. The material will be transported either within the main volume of ice or it may slowly descend to the bed eventually assisting scouring. If material falls onto the surface in the ablation zone it will usually remain on top of the glacier. There have been some interesting finds of material that have been incorporated and preserved by glacier ice. For example, in Siberia, an extinct woolly mammoth was found frozen inside a glacier, while in the Alps a prehistoric human was found.

The features formed by glacial erosion include large scale land-forms such as U-shaped valleys, with steep ridges and horns between the valleys. Here, ice masses flowed over the landscape, scouring it as they moved, leaving angular summits exposed that protruded above the ice. Glacial U-shaped valleys or troughs (note that river cut valleys are usually V-shaped) are mainly the product of abrasion, but rock fracturing and plucking downstream of smoothed obstacles are also important. While most glaciers follow existing river valleys, they act to deepen, widen and straighten them. Lakes often form in the eroded valley left by a glacier and these lakes slowly infill with sediment over time. If the glacier has eroded the valley to below sea level then the valley may get flooded after the ice has melted to form a **fjord**. Some infilling of the valley bottom occurs after ice melt as the gravels produced by meltwater downstream of the retreating glacier, combined with lake sediments, act to flatten the floor of the valley, assisting the production of the U-shape.

Often side valleys to the main glacial valley are truncated, leaving **hanging valleys**. These hanging valleys were occupied by tributary glaciers feeding into the main glacier. The landscape of Yosemite in California is a good example where there are plentiful hanging valleys often with large waterfalls tumbling over them for several hundred metres down vertical cliffs into the main valley. Small glaciers often form dome shaped depressions, known as **corries** or **cirques**, near the top of mountains. Once the ice has retreated the depression can form a small lake often called a **tarn**. The cirques are essentially small forms of hanging valleys. Glacial valleys and cirques may extend backwards particularly as freeze-thaw activity is intense, resulting in rock shattering. Two cirques that cut backwards into each other may leave a narrow ridge between them known as an **arête**. If three or more cirques cut back together they may leave a pyramidal peak or horn such as Mont Blanc in eastern France. Sometimes weathering or erosion of part of an arête can produce a dip in it known as a **col**. These often form low points in mountain ridges that humans have turned into routes to get across the mountains, as is the case for many Alpine passes.

On a smaller scale there are smoothed **whaleback forms**, which range from ten metres to hundreds of metres in length, which are the product of abrasion across the surface of an obstruction. The rock may be particularly resistant and so the glacier was unable to abrade it fully thereby leaving a smoothed mound orientated in the direction of glacier flow. However, many of these small mounds that are the product of erosion are not smooth all over. **Stoss–and–lee forms** are streamlined features that have a gently sloped, glacially smoothed upstream side and a steeper, plucked, downstream side. They are much more common than whaleback forms and occur when the glacier flows over the obstacle smoothing the upslope side. The stoss side is often scratched with grooves (**striations**) from sediment that has abraded the rock as it passed over it. On the downslope side of these forms, bedrock fracturing, loosening and displacement can occur and the fragments can be incorporated into the ice. Therefore, the downslope side of stoss-and-lee forms tends to be rough. Small stoss-and-lee forms are often called **roche mountonées**. These landforms tell us that there was warm ice in the glacier that once existed at that location. Roche mountonées are typically a few metres in height and tens of metres in length. There are many remarkable undulating landscapes where there are hundreds of roche mountonées spread over large areas. **Crag and tail features** form where resistant rock leaves small mounds protruding above the surrounding surface but on the downslope side the feature is formed from deposited sediment behind the obstruction. The setting for Edinburgh Castle, in Scotland, is a crag and tail feature.

The deposition of eroded material from glaciation has also produced important landforms. The glacier itself can produce the landform, or the landform may be left when the ice in the glacier melts. Finally, the meltwater can produce depositional features quite some distance from the glacier itself. Depositional landforms associated with glaciers are not as dramatic as those produced by erosion. Nevertheless, glacial deposits cover around 75 per cent of the mid-latitude land mass and 8 per cent of the Earth's surface. Features may be formed by the direct action of ice such as **moraines** which are linear mounds of sediment. Push moraines form when a glacier forces itself into sediment at the front of the glacier raising it into a ridge. These moraines can mark the maximum extent of glaciers in the past, helping us to map the extent of former ice masses. Dump moraines are formed at the front of a glacier where material being transported through the glacier is eventually deposited at its front end as the ice melts. There can be many of these left as a glacier retreats. Lateral moraines are those that have formed on the glacier surface collecting rockfall from the cliffs above. As the glacier moves, the intermittent rockfall debris appears as a linear feature on the glacier surface. Sometimes there can be massive boulders that are transported this way and once they are deposited a long distance from their original source they are often known as **erratics**. Hummocky moraines occur where there is a deposit of material from inside the glacier or on top of the glacier as the glacier melts.

Moraines are formed by the action of ice whereas **eskers** are features of water. Eskers are snaking ridges of sand and gravel that are thought to form mainly in R-channels at the glacier bed. They can be 20 to 30 metres high and up to 500 kilometres in length. Eskers may flow uphill as well as down, simply representing the fingerprint of an internal channel system within the ice mass. **Drumlins** are streamlined mounds of sediment, sometimes with a rock core, aligned in the direction of ice flow typically with a blunt end upstream and a more pointed end downstream. They often occur in large numbers across an area, such as the 10,000 or so that occur in west central New York State. There is considerable debate as to the mechanisms that cause drumlin formation, including that they were formed during really large flood events.

The outwash plain downstream of a glacier is often rich in gravels and boulders, and the meltwater system also contains a lot of rock flour and has a milky colour. Braided river systems are common in these environments. When large blocks of ice get swept down the river system and are deposited and buried they can later form **kettle holes**. The slowly melting ice leaves a depression in the surface where elsewhere in the surrounding landscape sediment had been deposited by the river system. On the other hand, braided rivers may deposit sediments that are several hundred metres thick in some locations, masking underlying features.

Permafrost

Permafrost refers to soil or bedrock that is frozen for longer than two years. It covers around 25 per cent of the land surface, although this is mainly in the northern hemisphere where there are significant land masses at high latitudes. More than half of the land area of both Russia and Canada, the world's two largest countries, is covered with permafrost. Permafrost usually consists of soil which contains ice within its pore spaces. Most of the Earth's permafrost is very susceptible to climate change because it exists at temperatures just a few degrees below 0°C.

At high latitudes, but outside of ice sheets and glaciers, permafrost tends to be continuous and often extends several hundred metres in depth. At slightly lower latitudes the permafrost may not be continuous and is also thinner, perhaps just being a few metres in depth. Continuous permafrost moves to discontinuous permafrost at a mean annual air temperature of around −6° to −8°C while discontinuous permafrost extends to a mean annual temperature of around −1°C within continental interiors. Close to the surface there is an annual melt and freeze cycle in both continuous and discontinuous permafrost. Temperatures measured in permafrost are typically lowest close to the ground surface, excluding the **active layer**, and increase with depth, reaching the melting point at the base of the permafrost.

The near-surface layer in which thawing and freezing occurs is called the active layer. This layer can be a shallow few centimetres in continuous permafrost to several metres deep in discontinuous permafrost zone. Freezing of water causes an expansion of 9 per cent in volume and so freeze-thaw processes in the active layer cause movements of the ground surface. In fact this can lead to major problems for infrastructure such as roads, pipelines and buildings which, if not properly engineered to cope with the conditions, will collapse, buckle and subside. Human-occupied buildings transmit heat to the ground and can melt permafrost, leading to subsidence. Therefore, buildings must be well insulated at ground level or even have refrigeration units to cool the ground although this can be prohibitively expensive. Ensuring that buildings are anchored to deep piles that are embedded to the bedrock is important and this means that the building could be supported off the ground if the ground subsides. The trans-Alaskan oil pipeline is a classic example of engineering which takes account of permafrost. The pipeline carries oil at 65°C across 1,285 kilometres of permafrost terrain. The hot temperature of the pipeline would thaw the ground if the pipe were buried or at the ground surface and this would result in subsidence and damage to the pipeline. Therefore, it is elevated above the ground, for a large proportion of its route, upon racks. It is also deliberately built with bends in it to allow the pipeline to expand and contract, move sideways and vertically without cracking. The vertical supports for the pipe are also equipped with devices to cool the permafrost. Animals can cross the pipeline at certain locations where the pipe is buried with refrigeration units.

Landforms in periglacial regions

Periglacial environments are those that are cold and subject to intense frost action but which are non-glacial. Permafrost regions are periglacial but permafrost is not a pre-requisite for a periglacial region. Many periglacial landforms are related to the freezing of water in soils and sediments. Frost heaving and thrusting are the vertical and horizontal movement of sediment due to the formation of ice. Heaving generally dominates because ice crystallisation tends to happen in a direction parallel to the temperature gradient and in soil this is usually parallel to the ground surface. Frost heaving moves masses of soil and may even push stones upwards to the surface. This may happen as the stone and surrounding soil gets pushed up during freezing in the active layer. When melting occurs in the summer, the finer sediments settle back down filling in the gap below the stone and supporting it. Additionally, soil water flowing around the stone may flow into the pore spaces below it and when this water freezes push the stone up once more. This leaves the stone a little higher in the soil profile each year. Over long periods this results in a net movement of stones to the surface.

The types of mass movement described in the section earlier in this chapter on weathering and erosion occur in periglacial areas. However, **frost creep** and **solifluction** are important in these environments. Frost creep occurs when sediment is pushed upwards on a slope during freezing, as part of frost heave, but then gravity pulls the sediment in a downslope direction when it melts and lowers so that over many years there is a net movement downslope. Frost creep will operate in conjunction with solifluction. Solifluction is the slow downslope movement of saturated soil in a very slow flow. It is more exaggerated in periglacial environments where this process in known as **gelifluction**.

Water can either freeze within pore spaces between solid particles of soil or sediment, or it can migrate to form discrete masses of ice known as **segregated ice**. Coarse gravels and sands are highly permeable but because pore spaces are large there is little 'suction' potential (see section on soils in Chapter 3) and so they do not retain much water. Finer soils such as clay have low permeability but high water retention potential. This means that soils with intermediate grain sizes, such as silt, have the greatest potential to form segregated ice within the ground. Segregated ice may form lenses or form bands. Where the bands are thick, sometimes up to several metres, they are known as **massive ice**. A large body of experimental research has shown that liquid films can coat ice surfaces even when the temperature is below the pressure-melting point. These films provide fluid conduits that supply the growth of segregated ice. Hence there is slow movement of very thin films (a few hundreds of a thousandth of a metre in thickness) of water that coats the ice and enlarges it. The movement of these thin films occurs from the sediment adjacent to the ice through several quite complex mechanisms including molecular attraction which is more powerful than the forces holding the water within the pore spaces. This then leaves relatively empty pore spaces near the segregated ice which become filled by meltwater seeping into the soil from above during the summer through suction processes, as described in the section on soils in Chapter 3. This means the segregated ice can slowly grow.

Frost cracking can occur by fracturing the ground as it contracts at very cold temperatures. While there is expansion when water freezes, as the temperature gets very low, ice and sediment contract in volume. This is probably a major factor in the formation of many polygonal crack features seen in the periglacial landscape. Frost crack polygons are usually 5 to 30 metres across and develop best when there is no insulating snow cover. These features cover large areas in North America and Siberia. Water entering the crack can later freeze and then these ice films expand over several years of seasonal melting and freezing to become **ice wedges**. These ground ice features can be 4 metres deep and 2 metres wide forming a V-shaped ice wedge in the enlarged crack. Therefore, vertical ice wedges in the sediment often accompany patterned frost crack polygons on the surface.

In addition to crack polygons there are often regular geometric patterns of stones or topography in periglacial areas. These can be grouped into circles, nets, polygons, steps and stripes. These are amazing features in the landscape and almost look like humans have made patterns by sorting stones and vegetation into neat shapes. Circles, nets and polygons are common on flat surfaces whereas steps and stripes occur on slopes between 5° and 30°. On steeper slopes, mass movement becomes a more dominant process destroying patterned features. Sorted stone circles typically have fine material in the centre of an area of lowered relief and larger sediment forms a higher perimeter. Sorted stripes look rather like a recently ploughed field with ridges and furrows consisting of alternating stripes of coarse and finer material. There are several mechanisms hypothesised for these features linked to heave and thrust processes and the possibility of mini convection circulation cells operating within the sediment near the surface. In the daytime, especially in summer, saturated soil near the surface warms while the soil below remains frozen. Water at the surface (at say 1° or 2°C) is slightly more dense than the thawing water at 0°C below (water is densest at 4°C). Thus the more dense water sinks and forces the less dense water upwards. Therefore, a small convection cell can form (think of the ocean or atmospheric circulation cells described in Chapter 1). It is thought that the edges of the convection cell are associated with the surface sorted features as soil particles may move with the water. However, the exact processes are not yet clear.

At a larger scale, **pingos** can form. These are ice-cored mounds up to 60 metres high and 500 metres in length. The mounds contain some segregated ice and a lens of massive ice. The top of the mound often becomes cracked as the ice core within the pingo grows larger, forcing the ground surface upwards. Hydrostatic pingos are caused by the doming of frozen ground as a result of the freezing of water and the growth of permafrost beneath a former lake or other water body. The features are usually isolated landforms predominantly in areas of low relief. Pingos that form over drained lakes are usually circular in shape, whereas pingos over old river channels may be linear in form. Hydraulic pingos form most commonly in discontinuous permafrost regions at the foot of slopes and are usually circular or elliptical in shape, and result from the inflow and freezing of groundwater seeping from upslope.

If segregated ice melts at any point then there can be a large volume of excess water and subsidence. This can lead to a terrain which is almost impassable, especially in summer, with lots of depressions, many of which are filled with water. This terrain, consisting of small irregularly shaped thaw lakes and depressions, is called **thermokarst**.

Periglacial mass movement on slopes can result in landforms such as **protalus ramparts** which are linear mounds of coarse sediment that form a small distance from the base of a slope. When a rockfall occurs boulders may slide across snow at the foot of the slope coming to rest just beyond the edge of the snow. **Ploughing boulders** move slowly downslope, leaving a depression on the hillslope indicating its path and forming a small bulge of sediment downslope of the boulder. The movement of the boulders, typically a few millimetres per year, is thought to occur because of the different thermal conditions beneath the boulder compared with its surroundings. Larger forms of ploughing boulders occur when a whole mass of rock and sediment moves downslope, a little like a glacier. These tongue-shaped **rock glaciers** have a steep front and usually descend from cirques that they have created through downslope movement of the angular debris. Ice within the pore spaces assists the flow processes.

While there are no perfectly symmetrical valleys, periglacial valleys can often show a distinct asymmetry. Areas which are no longer periglacial often have asymmetric valleys that are relicts from former periglacial times. The asymmetry can be caused by south-facing slopes having longer exposure to the Sun's energy in the northern hemisphere and this promotes prolonged thawing, more frequent freeze-thaw, as there are more days and nights with freeze-thaw conditions, more melting and more rapid mass movements. Therefore south-facing slopes in the northern hemisphere and north-facing slopes in the southern hemisphere will experience a quicker reduction in slope angle while the deposited material at the bottom pushes the stream toward the opposite facing slope undercutting it and keeping it steep.

Impacts of recent climate change

Evidence gathered by the IPCC has shown that ice masses and permafrost coverage is declining rapidly. Satellite information collected since 1978 shows that Arctic sea-ice extent has decreased at a rate of 2.7 per cent per decade, with larger decreases in summer of 7.4 per cent per decade. There are even predictions that the Arctic might be free of sea-ice in the summers at the North Pole within the next few decades. Mountain glaciers and snow cover on average (but not everywhere) have declined in both hemispheres. The maximum area of permafrost has decreased by about 7 per cent since 1900 while the temperatures at the top of the permafrost layer have generally increased since the 1980s in the Arctic by up to 3°C. While permafrost melt produces more thermokarst landscapes and adds to problems of ground subsidence and infrastructure damage, melting permafrost may also release more stored methane and carbon dioxide into the atmosphere as the soil organic matter is warmed and can be decomposed. On the other hand this may be balanced by faster plant growth that could take up additional carbon dioxide. The full impacts of permafrost melt due to climate change are, as yet, uncertain.

SUMMARY

- Natural water pathways, climate, vegetation cover, soil type, topography and management of the landscape influences how the chemistry of precipitation is modified as it moves through the landscape into rivers, lakes and deep groundwater.
- The pathways that water takes through and over soils and rocks influence the response of river flow to precipitation events.
- Overland flow production by infiltration and saturation-excess mechanisms generally results in shorter lag times and higher discharge peaks in the river than in deep throughflow and groundwater dominated systems. However, the exact outcome can depend on soil and bedrock type, topography, vegetation cover and climatic conditions.
- Flooding and drought susceptibility have been heavily modified by human action causing changes in water flowpaths, water storage and therefore in river flow.
- Coastal areas are dominated by wave and tidal processes that drive weathering and sediment movement.
- Coastal landforms can be characterised by wave-dominated features such as beaches, tide-dominated features such as estuaries and river-dominated features such as deltas.
- Coastal management must incorporate understanding of weathering and sediment transport processes because it is a tightly balanced system. Stopping natural sediment movements in one location on the coast may cause additional erosion and major coastal problems a little further along the coastline.
- Ice sheets and glaciers erode dramatic new landforms such as U-shaped valleys, horns and arêtes. They also create deposits that form other more subdued landforms such as moraines, stoss-and-lee forms and drumlin fields. These features can be seen in areas that are no longer glaciated but provide evidence of former colder climates.
- Permafrost is frozen ground which covers 25 per cent of the Earth's land surface.
- An active layer of melt near the surface operates in the summer in permafrost areas meaning that the ground periodically subsides.
- Buildings and infrastructure have to be carefully designed in permafrost areas to cope with seasonal melt of the upper active soil layer

and to prevent additional melt and subsidence of the wider permafrost caused by heating of the ground surface by infrastructure.

- Freezing water within sediment can form into large blocks of ice. Some of these can cause the surface to rise tens of metres, forming pingos.
- Frost action produces cracking, heave and thrust processes within periglacial areas producing landforms such as polygons, stone circles, stripes, asymmetric valleys and thermokarst.

FURTHER READING

On general topics covered by this chapter:

Holden, J. (ed.) (2008) *An Introduction to Physical Geography and the Environment*, second edition, Harlow: Pearson Education. A textbook with expert contributors providing more in depth material on all of the topics covered in this chapter.

Summerfield, M.A. (1991) *Global Geomorphology*, Harlow: Pearson Education. While this book is now quite old it is still very relevant and is a detailed text for those who want to pursue geomorphological topics in a lot more depth. There is excellent use of illustrations and examples.

On water:

Gupta, A. (2007) *Large Rivers: Geomorphology and Management*, Chichester: John Wiley and Sons. A book filled with case studies on river processes and management.

Lehr, J.H. and Keeley, J. (eds) (2005) *Water Encyclopedia*, New York: John Wiley and Sons. The five volumes of this work contain brilliantly detailed and clear explanations of water processes written by experts from around the world providing something to delve into if you want further explanation of how soil and river processes operate.

Price, M. (2002) *Introducing Groundwater*, second edition, Cheltenham: Nelson Thornes. This book provides very clear coverage of practical techniques for assessing groundwater and understanding groundwater movement.

Shaw, E.M., Beven, K.J., Chappell, N.A. and Lamb, R. (2010) *Hydrology in Practice*, fourth edition, Abingdon: Taylor and Francis. A good book aiming to provide a more detailed understanding of hydrological processes, measurement and modelling.

On coasts:

Davidson-Arnott, R. (2009) *Introduction to Coastal Processes and Geomorphology*, Cambridge: Cambridge University Press. This book covers the key topics and debates through a process-based approach.

Masselink, G. and Hughes, M.G. (2003) *An Introduction to Coastal Processes and Geomorphology*, London: Edward Arnold. This text provides excellent coverage of coastal topics.

On ice:

Benn, D. and Evans, D.J.A. (2010) *Glaciers and Glaciation*, second edition, London: Hodder Education. This is a very popular book with students who want to learn more about glacial processes and features.

French, H.M. (2007) *The Periglacial Environment*, third edition, Chichester: John Wiley and Sons. A clearly written and detailed textbook with diagrams and examples on periglacial features and permafrost.

Hambrey, M.J. (1994) *Glacial Environments*, London: UCL Press. A thorough treatment of glacial processes and landforms is provided by this book.

5

THE GEOGRAPHY OF ECOSYSTEMS

Biogeography is the term used to describe the geography of the biological world. It involves the study of the distribution and patterns of life on Earth and of the underlying processes that result in these patterns. It is biogeography which will be covered by this chapter. The biosphere is the biological part of the Earth which incorporates the Earth's surface and a shallow layer below it, the oceans and the lower atmosphere. Within the biosphere there exist many ecosystems. Ecosystems are biological communities and the physical environment that sustains them where energy and nutrient cycles link the organic and mineral components of the biosphere.

THE BIOSPHERE

The biosphere is characterised by large- and small-scale energy flows and cycles of nutrients. The Earth's biosphere is not the same throughout, but has patterns of distinctive regions at all scales from the global to the tiny. Variations within the biosphere may result from factors including climate, geology, soil type, human action and biotic processes.

Ecological variables

Light

Photosynthesis by green plants captures carbon and oxygen from the atmosphere and water from either the atmosphere or the Earth's surface and combines these to produce complex carbohydrates, which are the building blocks of all life. The energy for photosynthesis comes from the Sun. Around one sixth of the light energy absorbed by a green plant is used for photosynthesis while the rest is converted into chemical or potential food energy of the plant tissues. This energy can be used by other organisms consuming the plant tissue. The energy is released as heat through respiration in plants and animals which consume oxygen and release carbon dioxide. Green plants therefore need light, and the more light they have the more growth can be expected. Thus faster growth would be expected in the tropics and slower growth at the poles. The majority of plant species (C_3 plants) are found to fix carbon dioxide into '3-carbon' compounds known as triose phosphates. However, some other species (C_4 plants) make a '4-carbon' compound instead, known as oxaloacetic acid. C_4 plants have only evolved recently with many being grasses, sedges and a few herbs and shrubs. C_4 plants have an advantage over C_3 plants in that they can utilise high levels of solar radiation effectively, use water more efficiently and are more drought tolerant. They may therefore be favoured by climate change over the next few centuries and are among the fast growing crops of the world, such as maize, sorghum, millet and sugar cane. Note that it has only recently been discovered that light is not a prerequisite for life and ecosystem development. Instead of photosynthesis, some deep ocean systems have developed **chemosynthesis** (see Box 5.1).

Box 5.1 Deep ocean life in the dark

At mid-ocean ridges (see Chapter 3) over 2 kilometres below the ocean surface, where light does not penetrate, there is life. This life is sustained by energy, not from sunlight, but from hot vents in the ocean floor. These hot vents emit water and many dissolved chemicals and black particles. Surrounding the vents there are large communities of animals including tube worms several metres long and blind shrimps. Unlike photosynthe-sising green plants at the top of the ocean or on land, bacteria around the vents gain their energy from the chemicals released by the vents such as hydrogen sulphide or methane, in a process known as chemosynthesis. These bacteria are then grazed upon by other creatures, creating a **food chain**. Some of the bacteria even live in the shells of other creatures. The deep ocean creatures have to avoid being boiled by temperatures from the vents which can be over 300°C. The discovery of these ecosystems means that light is no longer considered to be prerequisite for life.

Temperature

The optimum condition for growth and photosynthesis for most (but not all) plants is between 10°C and 30°C. Seasonal patterns of temperature are important as the growing season for most plants creates a baseline of food provision for other creatures. The growing season is particularly important for **herbivores** (animals that just eat plants) which must adapt to the changing availability of food resources through the seasons. They often do this by becoming dormant (e.g. hibernation) for part of the year or by migrating.

Moisture availability

Moisture availability is mainly linked to rainfall regimes. However, temperature and the ability of soils and rocks to store water that is available for biological use are also important. Soil type, geology, slope and altitude are often crucial in determining areas of increased moisture for plants and animals. All the important plant reactions take place within water. For plants on land, water also supports their structure and without it they wilt.

Other climatic factors

Humidity can control photosynthesis, which often switches off in dry air. Wind can influence local temperatures. If there are strong prevailing winds in a particular location then only strong plants able to withstand windy conditions may grow there.

Geological factors

The movement of plates across the surface of the Earth (see Chapter 3) has provided opportunities for species to spread or for barriers to form such as chains of mountains or opening of oceans. For example, there are large differences in fauna and flora between the islands of Bali and Lombok, 30 kilometres away. Those of Bali are related closely to those on the larger islands of Java and Sumatra to the north. Those on Lombok are more like those on New Guinea and Australia to the south. Between Bali and Lombok there is an ocean trench which has separated the plates for over 200 million years prevented any connecting land from joining. The islands to the north of the trench were formed from the Asian continent and the islands to the south were originally part of the Australian continent. The role of plate tectonics is of fundamental importance to explaining many spatial differences in plant and animal distributions. Other geological factors include soils, which are important for controlling available water for plants and provision of nutrients, and topography, which influences receipt of the Sun's energy, local climate, hydrology and soils.

Biotic factors

Competition for light, nutrients, water and living space, the ability to adapt and migrate and the presence or absence of predators and prey are important components that may result in differences within the biosphere. Competition arises in situations such as at a drinking hole in a semi-arid area. Herbivores and other species may need to wait until the carnivores have left before getting a drink of water. Another type of competition occurs when there is a limited amount of food for the individuals of one species and this may lead to exclusion of weaker individuals and the survival of the fittest, who are able to gather enough food. The **ecological niche** is the basis of most ecological patterns in the biosphere. Where there are no competitors for any of the resources required by an individual or species, the organism can occupy the ideal conditions to which it is adapted. However, because of competition, species usually have to occupy a niche that is the result of competitive interaction between several species attracted to the resources. Competition tends to be strongest between similar species, since their ecological niches are likely to overlap. The species able to survive on the lowest amount of the limiting resource will be better off.

Another biotic factor that produces geographical patterns is the isolation of groups of organisms, perhaps through plate tectonics. The biotic component of this isolation is the lack of breeding of the species with the larger population. This means that a wider mixing of genes does not occur and so adaptations or changes to a species can more quickly develop. This can lead to the evolution of new species (**speciation**) such as with the Galapagos finches studied by Charles Darwin.

Organisms can also vary widely in how well they can move or spread into different areas and this is an important determinant on the combinations of species to be found in an area or how the system might respond to environmental change. All organisms disperse their offspring. Some plants disperse millions of seeds but it may be that only a few will survive. Others only release a few seeds under certain favourable conditions for growth. These biotic processes help shape geographical patterns within the biosphere.

The ecosystem

Ecosystems vary from huge rainforests to individual rocks. Ecosystems involve the flow of energy and nutrients within the cycle of life. This means that changes to one part of the ecosystem will affect other parts of the ecosystem. Ecosystems can be divided into several energy levels known as **trophic levels**. The lower level photosynthesisers use the Sun's energy, nutrients and water within the soil to produce organic matter. These are the primary producers. This plant matter is then eaten by herbivores who occupy the second trophic level. Some of these may be eaten by carnivores (meat-eaters) occupying the third trophic level while some of these carnivores may be eaten by other carnivores occupying the fourth trophic level. During this whole process, waste is generated and this may be recycled back into the soil. When organisms die, their remains will probably form the diet of decomposers (e.g. maggots or fungi) who transform the litter to humus (see Chapter 3). This process releases the last of the energy as heat (decomposing compost heaps are often very warm). The decomposition process is important in the cycling of nutrients that have been passed through the food chain such as nitrogen and phosphorus.

Of course, most ecosystems are more complex than the trophic levels that have just been described but the principles remain the same. Much of the energy gained by herbivores in consuming the primary producers is used in motion, digestion, respiration and so on. Therefore, perhaps only 10 per cent of the energy is passed on from one trophic level to the next. This means that a high trophic level consumer requires a lot of primary production to support it.

For a human to eat 1 kilogram of wild salmon would require 1,000 kilograms of phytoplankton to have been produced. In addition to energy transfers, material is transferred through an ecosystem. The trophic system means that certain toxins that might be present originally in low concentrations can become concentrated in the high trophic level consumers under certain conditions (see Box 5.2). Two further important examples of material transfers are through the nitrogen and phosphorus cycles which are described below.

Phosphorus

Rock weathering enables phosphorus to be released in a soluble form in water solution. It then becomes available within soils or water bodies for uptake by plants before being consumed by organisms at higher trophic levels. Excretion or death means that phosphorus can be taken up by decomposers and the phosphorus becomes part of the soil of water solution again. This cycle can continue over long periods. On a larger scale phosphorus may be washed off the land and into the deep ocean sediments. Over time this may form a sedimentary rock which then later reaches the surface, weathers and becomes part of the water solution in soil.

Nitrogen

Nitrogen is essential for life. Nitrogen exists in large quantities in the atmosphere. Very small amounts of nitrogen gas react with oxygen during lightning to form nitric oxide, which eventually reaches the ground as nitrate. More importantly, soil bacteria fix nitrogen to make reactive forms of ammonium and nitrate which can be used by plants. Some of these nitrogen-fixing bacteria form close relationships with plants such as legumes (e.g. clover). When used in crop rotations these nitrogen-fixing plants can help fertilise the soil from the atmosphere, ready for the next crop. Ammonium and nitrate are taken up by plants to make protein and other parts of the plant matter. Herbivores then eat plant material to obtain their protein and therefore their useable nitrogen. Once passed through the trophic system, dead plants and animals decompose and some of the nitrogen is acted upon by denitrifying bacteria. Nitrogen is thereby returned to the atmosphere.

Humans have disturbed the phosphorus and nitrogen cycles not only by developing fertilisers but also by changing the properties of soils, accelerating soil erosion and also impacting the ability of phosphorus and nitrogen to become available to plants in soil water. Furthermore, humans have also produced more nitrous oxide from industrial activity and vehicle emissions. This has resulted in enhanced nitrogen deposition from the atmosphere in rainwater, for example.

Box 5.2 Bioaccumulation

Through the cycling processes of the ecosystem there may be locations where some chemicals become concentrated and are available for uptake by organisms which are not desirable. When producers and consumers take up nutrients they may also take up other materials. As a result, toxins may accumulate in specific parts of an ecosystem, usually in the higher levels of food chains. For example, a contaminant such as mercury can build up in the sediment of the seabed and then be taken up by mussels. Each mussel may only contain a small amount of mercury. However, as small fish eat lots of mussels the mercury becomes more concentrated within the small fish. As predators eat the small fish then the mercury can accumulate to high levels within them. This can have health consequences for humans who may eat the larger fish such as tuna and swordfish.

Ecosystems are dynamic and are constantly adjusting to changing environmental conditions or disturbance. **Succession** occurs when older groups of plants and animals are replaced by more complex groups. Primary succession may begin on a bare rock or disturbed site. Lichens, mosses and ferns are often part of the early colonisers. Through inputs of organic matter and continued weathering and/or shelter, these plants help alter the site conditions and thereby make them more suitable for other plants and animals to colonise. These secondary colonisers in turn build up the complexity of the site and alter conditions further and so the ecosystem changes through time. Succession on ponds and lakes can result in the eventual infilling of the lake by organic matter and other debris.

Island biogeography

The study of isolated islands has yielded much useful understanding about how biogeographical processes operate. Islands have clear boundaries and their isolation and lack of many external factors simplify the system and make it easier to understand. Classic island biogeography theory examines the balances between rates of immigration of new species to an island and rates of extinction on that island. If immigration rates are high and extinction rates are low then the island should be rich in species. The species richness should be related to the size of the island and how far it is away from another land mass. If an island is a short distance from a major land mass then there will be more opportunities for new species to arrive. Larger islands are more likely to have a wider variety of habitats and so might be able to support a larger range of species. The rate of extinction of species inhabiting a new island starts off low, since competition for resources would be low. As the number of species increases then the pressure on resources also increases and so the rate of extinction rises over time. If an island is created by severance from the main continent then the island might start off with high species richness. However, as the island has restricted resources then extinction rates will increase at first resulting in a decline of species richness.

Island biogeography theory has also been used to help understand best management practice for conservation. For example, there have been questions about whether it is better to conserve one large area within a landscape or conserve several smaller areas. One large area might be more species rich and work well for conservation, particularly if there are large scale migratory routes through that landscape. On the other hand, while lots of small patches might mean there is less species richness within those patches, the chance of complete loss of that ecosystem (and extinction of species) is reduced because there are lots of patches with replicated ecosystems.

THE BIOMES

Biomes are global areas containing major terrestrial vegetation communities with similarities between the dominant plants and

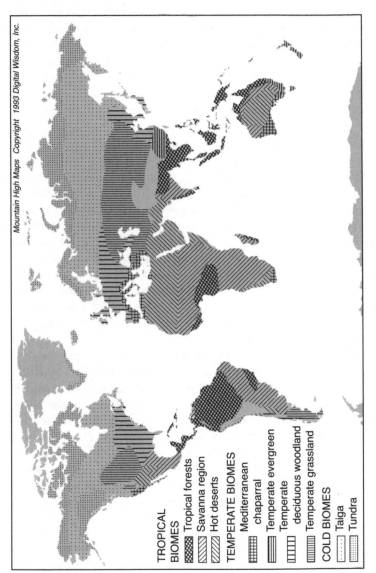

TROPICAL
BIOMES
▨ Tropical forests
▧ Savanna region
▨ Hot deserts

TEMPERATE BIOMES
▦ Mediterranean
   chaparral
▥ Temperate evergreen
▥ Temperate
   deciduous woodland
▦ Temperate grassland

COLD BIOMES
☐ Taiga
▧ Tundra

Figure 5.1 A map of the terrestrial biomes. Unshaded areas are covered with ice or are dominated by mountainous biomes.

Mountain High Maps Copyright 1993 Digital Wisdom, Inc.

characteristic animal communities. The location of the major global biomes is shown in Figure 5.1. Climate is the main driver of the location of biomes, with temperature and precipitation regimes being crucial. The ocean is not really divided into biomes but there are roughly horizontal layers which support typical plants and animals.

Cold biomes

The cold biomes consist of the area including the forested taiga to the treeless tundra. Note from Figure 5.1 that these biomes are restricted in the southern hemisphere due to a lack of ice-free land at high latitudes. The cold biomes are therefore often called the Boreal zone (which means the northern zone).

The taiga extends north from where the monthly average temperature of 10°C occurs for less than five months of the year up to where only one month has an average temperature above 10°C. The growing season is therefore short and soils are often thin because large areas have been eroded by former ice sheets and soil development is slow due to cold temperatures. The lack of soil animals means that decomposition is limited and there is an acidic leaf litter. In many areas, large expanses of peat have developed due to impeded drainage caused by the underlying fine glacial deposits.

The taiga forests of Europe and Asia consist mainly of coniferous trees such as Norway Spruce and Scots Pine in the west but there is more deciduous larch in the east. In North America, Lodgepole Pine and Alpine Fir dominate in the west and White Spruce, Black Spruce and Balsam Fir in the east. Tree growth rates are slow, especially in the colder areas with a height gain of perhaps 15 centimetres per decade, but evergreen species are able to photosynthesise as soon as conditions are right rather than having to wait for leaves to develop. The forest structure is simple with an evergreen canopy and then ground level tufted grasses, mosses, lichens and heathland plants such as bilberry. Tree cover is not dense everywhere and there are more open areas especially where it is colder or where soils are very thin. Active layer processes in permafrost can cause ground instability that can fell individual trees. Where tree cover declines the ground is often covered with lichens. Caribou (North America) and reindeer (Europe) migrate through the taiga forest. Smaller mammals, such as weasels, tend to be camouflaged with fur that changes colour in winter.

Wildfire is an important feature of the taiga. Fire is helpful in nutrient cycling, speeding up an otherwise very slow process, removing a thick, acid, litter layer (which may have been keeping more of the ground frozen for longer). Lightning strikes can initiate large fires perhaps at 200 year intervals. These disturbances clear land to ensure there is a good species mix but species are also adapted to fire with many conifers, shrubs and herbs sprouting from roots, stumps and underground stems or having long seed dispersal periods (e.g. from pine cones).

There is a gradual transition between the southern taiga and tundra to the north with trees becoming sparser in colder conditions. Tundra is the rather flat, treeless zone between the taiga and the polar ice. These regions are harsh for life: summers where temperatures rise above freezing may only last a month or two; winters may see temperatures as low as −50°C. Strong, cold, dry winds are common, with little precipitation. Permafrost with a shallow active layer inhibits plants with deep roots and soil animals. The soils are shallow with acid litter above a gleyed horizon and then a permanently frozen layer. There is very low productivity and species richness with low growing, woody, herbaceous plants and mosses and lichens. Growth rates can be so slow and summers so short that some plants produce flower buds one summer ready for opening and pollination the next year. The low productivity means that these areas can be susceptible to human damage as recovery rates would be very slow indeed (e.g. from tyre tracks, mining etc.), leaving marks for hundreds of years. The low productivity also means that large areas of land are required to support any herbivores (e.g. rodents or migrating herds of reindeer and caribou) or carnivores (e.g. owls or foxes). Lemmings are important in the tundra zone as they increase the rate of nutrient cycling among the soil and plants and can even stimulate growth rates through their grazing action. They act as important prey for carnivores but the population of lemmings changes over cycles of a few years having knock-on effects for the rest of the ecosystem.

Temperate biomes

The temperate biomes can be split into Mediterranean chaparral, temperate grasslands, temperate deciduous forest and southern temperate evergreen forest. The Mediterranean biome is more than just the area around the Mediterranean Sea but can be found in California, west South Africa, central Chile and southern Australia. Mediterranean chaparral climates are warm all year but with low rainfall, summer drought and high evaporation rates. The Mediterranean biome is characterised by hard, tough-leaved plants, mixed woodland and scrubland adapted to growth in conditions of limited water availability. Many plants in Mediterranean climates are adapted to frequent natural fires. Trees have thick, smooth bark and deep roots from which new growth may occur. Many types of seeds only open after exposure to fire. Animals have adapted to drought and fire, often by being able to escape quickly (e.g. kangaroos, elk, goats and emus) or by burrowing (e.g. bobcats and rodents).

The temperate grasslands are found in extensive areas across central North America and central Eurasia. As the name suggests, the vegetation is dominated by grasses, usually perennial (the same plant surviving for year after year). There tends to be a long dry season and annual precipitation totals of usually less than 500 millimetres. The gently undulating landscapes familiar of regions such as the prairies and the steppes have often become growing areas for cereal crops. In fact in North America there are very few areas of natural tall or short grasslands left. Large herds of herbivores such as bison were typical but are now reduced through human action. Leaf surface area tends to be small in temperate grasslands to reduce transpiration. Lack of a protective upper canopy cover means that animals have had to develop speed to escape from predators (e.g. antelope and deer), a large size to reduce the chances of being attacked (e.g. elk and bison) or burrowing (e.g. voles). Frequent summer fires means that systems at or below ground level that allow survival, such as bulbs, rhizomes or tubers are important. Scrub and trees with suckers have developed in some wetter areas of temperate grasslands such as the South African veldt, or the tussock grassland of New Zealand and in Australia.

The temperate deciduous woodland biome is only found in the northern hemisphere. For similar areas in the southern hemisphere the woodland is evergreen. The differences may be related to plate tectonics where deciduous habitats only evolved in the north after the large land mass of **Gondwanaland** split up. There is a fairly short transitional zone between the deciduous forest and the more northerly Boreal forest biome. The climate in the deciduous forest biome is moist but temperate all year, with over four months having a mean temperature above 10°C. Well-mixed soils are typical with a rich soil fauna and the soil holding plenty of nutrients. In the best areas for the forest there are four main layers to the structure of the vegetation. The upper canopy at up to 30 metres high has broad, rounded tree tops. There is a shrub layer below around 5 metres in height, with a third layer of grasses and a ground layer of mosses and liverworts. The vegetation cover is seasonal with dramatic changes to the appearance of the system through the year. The loss of leaves in winter reduces transpiration and frost or snow damage. Flowering of trees is usually very early in the spring providing as much time as possible for fruits to develop. Some animals may hibernate during the more dormant period or burrow to avoid winter cold. Deer and bears were common but humans have modified these landscapes through hunting and deforestation.

The southern evergreen forest tends to have a similar climate to the deciduous forest in the north but generally there are two upper tree canopies and a shrub layer. As the forest is evergreen then ground level plants are less common as there are limited opportunities for good light to reach the forest floor.

Tropical biomes

The tropical biomes cover highly productive tropical rainforests, less productive savanna and low productivity hot deserts. Tropical forests tend to be consistently moist and warm, typically occurring where average annual temperatures are around 25°C with little seasonal variation, and where there is around 2,000 millimetres of regular rainfall each year. The regular water supply creates plentiful stream networks and major rivers such as the Amazon and Congo. Soils in tropical forests are deep, but relatively infertile since most of the nutrients are stored within the above-ground structure of organisms, although an amazing number of soil organisms also store biomass. The cycling of nutrients is extremely rapid so plant litter does not accumulate very deep. The tropical rainforests are lush with broad-leaf evergreen tree cover. The tallest trees tend to be narrow with few branches or leaves below the top of the canopy. These areas produce around 40 per cent of the world's land-based primary production and contain half of the world's faunal and floral species. The lack of strong seasonality means that fruit production and growth can continue across the forest all year and there is a dense leaf canopy as the plants compete for every last bit of light. Climbers and epiphytes (plants which grow above the ground surface using other plants for support and that are not rooted in the soil) are common. Lianas (a type of climbing vine) climb rapidly and will not usually form leaves until sufficient light is available. The rainforests contain a huge diversity of animals with many adaptations for climbing such as monkeys with strong tails, snakes and lizards. There is relatively little vegetation on the forest floor as it is so dark (only around 1 per cent of the light at the top of the canopy) but this provides room for ground-based large animals such as pigs, leopards and jaguars.

Savanna temperatures tend to be similar to those in rainforests but the long dry season (rainfall less than 250 millimetres per month for more than five months) means that the vegetation is seasonal in nature and is often adapted to be drought tolerant. The high rates of evaporation and transpiration mean that rainfall needs to be plentiful at these temperatures to achieve high productivity. A sparse tree cover allows the growth of grass and other ground flora. The structure varies across the landscape representing local differences in water and nutrient availability. Savanna plants are adapted to withstand fire and drought, such as the baobab which has thick bark, a short leaf season and a trunk that stores a lot of water from wet periods. Deep roots to capture water and thorns and spines to deter grazers mean that plants only need a few leaves. Fruiting of trees and other plants tends to be dominated by fire which can occur every few years. Fruits are dropped at the end of the fire into the soil. This soil is temporarily rich in nutrients from the deposited material left after the fire. Large animals are often found in the savanna, particularly in Africa, such as wildebeest, antelope, zebra, buffalo and elephant. Nocturnal animals have adapted to reduce water loss and hide from predators (e.g. aardvark).

Hot deserts do contain plants and animals. The soils are generally poor and lack cohesion though since the plant cover is extremely sparse and organic matter inputs are therefore low. Where water is concentrated, the habitat improves but most of the biomass is underground. Adaptations focus on maximising the conservation of available water; cacti store lots of water within their stems; leaves are often replaced by thorns. The creosote bush has a wide distribution of roots which put toxins into the soil to prevent other seeds, including those of the creosote, from germinating to ensure there is no competition for water. The woodiness of many desert plants helps stop the collapse of plant material during wilting. Grasses tend to be short and tufted to protect against drought and heat. Lengthy dormant seasons are common for plants. Seeds often only germinate when the conditions are wet, perhaps many years after they have been deposited. This can mean there is a sudden bloom of life during a wet period and then offspring are not seen until the next wet event. Animals are adapted to reduce moisture loss either by being nocturnal or only producing small amounts of urine. Some animals and plants are designed to capture dew, such as the Namib beetle.

Mountain biomes

Topography can be an important factor influencing biomes. High mountains usually exhibit strong contrasts in the ecosystem with altitude. The zones that can be identified vary widely between mountain regions but a typical classification may involve: a hill zone where the flora and fauna is related to that in the lowlands; a montane zone which contains species that are common to mountains but usually dominated by deciduous forest; a sub-alpine zone dominated by conifers with shrubs at the upper fringes; an alpine zone lacking in trees with vegetation that is short with grasses, sedges and plants with flat, spreading growth with reduced leaves and often with colourful flowers; and a snow zone which has sparse vegetation consisting mainly of mosses and lichens leading up to the permanently covered ice and snow zone.

HUMAN IMPACT

Humans have modified the biosphere by overexploitation of species, deforestation and extinctions, the distribution of species either accidental (rats on ships colonising islands being visited) or deliberate (introduction of new crops to an area), and the evolution of species through domestication of crops and animals, soil erosion and environmental pollution. Many ecosystems are complex and it is normal for complex systems to be more robust than simple systems when faced with change. Given the vast changes in climate experienced over the past 2.4 million years alone (see Chapter 2), it seems that the biosphere is able to accommodate vast fluctuations in climatic and environmental conditions and changes in the numbers and density of the species which it supports. However, humans have a large capacity to alter the ability of the ecosystem to cope with such stresses. Furthermore, if the system is highly interconnected then the removal of one key species could have major impacts. Species that are highly connected with the rest of the food web are those whose elimination is likely to be the worst for the ecosystem. These are known as **keystone species**. Prairie dogs are keystone species in temperate grasslands, which act as grazers, predators, prey and provide habitats for other species through burrowing.

Biodiversity

Biodiversity has various measures but essentially it is a term that describes the number and variety of species within an ecosystem. Global areas of high biodiversity usually result from lack of disturbance and lack of isolation. Regional and local patterns may result from short-term disturbances (fires which maintain overall ecosystem diversity) or habitat diversity. There is concern that human activities, including accelerated climate change, are causing a decline in the number of species. While 1.5 million species are currently known there may be twice this number yet to be discovered. Since the tropical forests hold around half of the world's species, their deforestation is of major concern. Over the past 400 years around 500 plants and 600 animals have become extinct that we know of, mainly due to human action. Natural extinction rates are normally around one mammal per 400 years and so the current rate of 36 mammals in the last 400 years far exceeds this. There may be many undiscovered species that are being driven to extinction as you read this. That is a cause for concern not only because it suggests we are damaging the ecosystems around us but also

because many plants have been found to have important pharma-
ceutical benefits for us and if they are 'lost before they are found'
then their medical uses will never be discovered. The elimination
of species by direct slaughter and over-killing by humans for food,
highly priced animal products such as ivory and whale oil, or to
remove pests has led to major modifications of the animal kingdom.
Slaughter by humans can be on an enormous scale. For example, of
the 60 million bison on the Great Plains in 1700 only 21 individuals remained by 1913.

Half of the Earth's forest cover has been removed by humans.
Deforestation commenced several thousand years ago, accelerated
by the development of the axe. The emergence of domestic livestock encouraged the clearing of land for agriculture, fuel needs
grew and protection from enemies was realised through removal of
hiding places in forests. Deforestation has progressed across different zones of the Earth and its current focus is now in the tropical
forests. Large amounts of carbon are stored in the biomass of the
tropical forests. Unlike in temperate forests, the nutrient store is
mainly above ground and so even if the trees are removed and used
for furniture (and hence the carbon is still on land) then the nutrients have been taken out of the forest and the ecosystem will be
severely depleted. It also seems that the tropical forests have helped
absorb some of our carbon dioxide emissions from fossil fuel
burning, acting as a buffer against climate change and thus their
removal also destroys this buffer (see Chapter 2).

It has been suggested that there are hotspots on the land surface
where biodiversity is particularly high. Work by Norman Myers
and colleagues has suggested that 44 per cent of all species of vas-
cular plants are found in 25 hotspots comprising only 1.4 per cent
of the land surface. Surprising areas are included as hotspots; the
natural vegetation of the tropical Andes seems to be the most
diverse hotspot. Even though its vegetation has been reduced to 25
per cent of its original extent by human action it still contains 6.7
per cent of all plant species in the world and 5.7 per cent of all ver-
tebrate animals. Myers and colleagues urge that these hotspot areas
be singled out for the attention of conservationists, to attempt to
protect them. It is also notable that climate change projections
suggest that all of these hotspots will undergo considerable warming
in the next 100 years.

Aliens and biosecurity

Humans are also agents of dispersal and distribution of species. The growth of long-distance transport has effectively bridged former barriers to movement for species. The deliberate introduction of wild animals from one region to another has often had unintended consequences on the ecosystem. Rabbits were introduced to Australia in 1787 and 1791 and, lacking predators, were able to grow their populations at enormous rates, damaging the native vegetation cover and becoming pests. Foxes were then introduced as the natural predator to the rabbit but instead preferred to prey on native marsupials and birds, devastating their populations. The grey squirrel introduced from North America to Europe is currently causing the displacement of the native red squirrel in Britain. Many species of plants and animals (e.g. rats) have been accidentally transported in vehicles, ships and planes and become alien invaders, sometimes spreading disease or driving other species to extinction. Deliberate and accidental introductions of species, which act as weed species without their natural controls of predators, alter local balances. Being careful to restrict the number of unwanted alien introductions is therefore an important activity at airports, ports and other borders as part of biosecurity. Ships, aircraft and mail are searched and sanitised as a matter of routine by many countries. Another routine procedure requires ships to empty water-ballast tanks mid-voyage.

Agriculture

Agriculture now dominates the landscape across many areas of the world. The area of agriculture is expected to enlarge by as much as 50 per cent as the world's population grows. The agricultural process will also consume more water which may restrict water availability for other ecosystems. Agricultural systems are ecosystems with managed inputs and outputs of energy and nutrients with controlled species diversity. Energy flows occur along simple routes. A higher proportion of the Sun's energy is made available to crop plants and then passed on more directly to humans or indirectly through livestock. This also means that a higher proportion of the primary production is exported from the system as the harvested product and therefore there is less organic matter and nutrients in the soil. Nutrients and organic matter are therefore added to the system by adding fertilisers or by having crop rotations with fallow years. Agricultural systems tend to have low biodiversity and are fairly simple. Selective herbicide, reinforced by the use of selective fertilising or manual or mechanised weeding, reduces diversity. Further impacts of agriculture include increases in soil erosion rates (see Chapter 3) and leaching of pesticides and fertilisers into watercourses causing aquatic ecosystems to be altered.

Humans have encouraged species evolution through agricultural processes. Normally evolution is slow. Variant individuals of a species only give rise to a new strain under conditions that favour their survival and allow them to maintain their variance by avoiding cross-fertilisation with the 'normal' members of the same species. This is quite rare. Also mixing two species to get a hybrid is also quite rare. This is because of cross-fertilisation back with the parent species dampens any variance. However, humans can create habitats that give variants a competitive advantage and indeed humans deliberately select plants and animals for domestication. The selection, planting and propagation of favoured plants, their variants and hybrids to suit human needs means that these plants grow well in the conditions provided for them, but many of them would not survive in the face of competition in the wild (e.g. they have lost their thorns, hairiness, toughness and so on which is good for human consumption but bad when it comes to surviving in the wild). In fact now, most crop plants lack the ability to reproduce or maintain themselves independently. Many food crops are entirely dependent on humans for their propagation. The banana is a sterile hybrid which produces attractive fruit but is unable to develop seeds necessary for its own propagation. It should also be noted that around 85 per cent of food supplied to humans today is derived from less than 20 plant species and there are concerns that these species may no longer be tough enough to withstand a new disease and so we should be diversifying our food sources. **Genetic modification** is another step along the route of evolutionary change that humans have fostered in their search for more productive food crops. Instead of cross-breeding or selective breeding of crops, genetic modification (often called GM) speeds up the process of modification by applying the changes that humans are seeking into their DNA by inserting or deleting genes. This provides a precise result rather than the less precise cross-breeding approach. Genetically modified food seems to be of concern to members of the public in some countries although the reasons for this are not entirely clear given the long history of domestication and selective breeding.

There are relatively few species of domesticated livestock and these tend to be herd animals which can survive on low nutrient vegetation and therefore use land that may be less suitable for crops. It seems that pigs and cattle have been living in proximity to humans since about 10,000 years ago and by 2,000 years ago pigs, cattle, sheep, goats and buffalo had all been domesticated and had evolved into breeding groups distinct from their wild ancestors. Selective breeding has brought about changes in the physical and physiological features of these creatures but more recently genetically modified pigs have been developed. Selective breeding really can result in fast changes to an animal species. For instance, the variety of domestic dogs we are familiar with such as poodles, Labradors, terriers, huskies, bulldogs, Alsatians and so on all originate from one single species – the wolf – by humans selectively breeding the animals only over the last 12,000 years.

Food consumption tends to be more protein rich in developed countries with a high proportion of meat and dairy products. Trends show that as developing countries progress, these demands also increase. This is problematic because it takes more energy, land and water to produce these foods. In fact, half of the agricultural land of the USA and Canada is currently used to provide plants for animal consumption. Therefore not only will population increase drive changes in arable production, but the wealth of nations will change the types of production as food preferences change too. Many areas are already stretched for water resources but an increased demand for meat and dairy products will increase water use dramatically with impacts on other parts of the natural system.

Urban ecosystems

Half of the world's population lives in urban locations. These urban locations, with specific domestic animals, pests and careful or careless planting, have resulted in urban ecosystems. These ecosystems include highly managed parks and gardens with deliberate introductions of species as well as abandoned land with habitats containing native and alien species. This means that habitat diversity in urban areas can be high. The various structures produce large and small habitats that are quite different and changes in buildings and land use means there is often change in habitats across different parts of the urban zone.

The warmer temperatures, resulting from the urban heat island effect (see Chapter 1), modified air flows and poorer air quality, compared to nearby rural areas, are components of urban ecosystems. Typically, urban areas produce more organic waste through sewage and foodstuffs than can be biologically decomposed and cycled back into the ecosystem. Some species have found the combination of resources in urban zones to be to their advantage. The rich nutrition provided by waste dumps attracts many birds and other species. House mice and house spiders, together with the brown rat are all part of the urban nutrient cycle. Released pets have sometimes colonised urban areas (e.g. parakeets) that are not native. Predators at higher trophic levels have also come into urban areas to take advantage of new prey available in the urban ecosystems. Indeed, agricultural change in rural areas has removed places of shelter, such as shrubs and woodland, and may have pushed some species into urban areas where the shelter is better.

Climate change

Research has shown that vegetation zones on mountains have been moving upwards and trees expanding from the taiga further north. Spring flowering has been occurring earlier and growing seasons have been lengthening. These changes have been taking place but it is not certain how the biosphere will respond to rapid climate change forecast to take place over the next 100 years (see Chapter 2). There are simply too many plant species to examine how each one individually might respond given its characteristics. So instead a number of models have been designed that predict how systems might respond using **plant functional types** (plants that share similar traits and similar in their association with environmental variables). These models seem to suggest that some tropical forests in south-east Asia, central America and the Amazon will become savanna and some of the Mediterranean chaparral will also become savanna. The models also suggest that evergreen forest will replace grasslands in parts of North America and Northern Europe. Of course the whole picture is quite complex because plants may grow better with more carbon dioxide in the atmosphere and also pollution that adds nitrogen to the atmosphere may provide more nitrates and ammonia for plants, enabling better growth in some systems.

Humans quite like to keep environments in a static form and often conservation of an environment tries to protect a status quo. However, climate change may mean that biomes have to shift and migration of plants and animals will occur. Humans who try to 'maintain the line' and keep an ecosystem operating in a less than ideal climate will struggle against a changing climate. Therefore, climate change will represent a significant challenge for conservation.

Conservation

Conservation can mean different things to different people. An important species to protect for some may be a pest to others. Conservation motives can include ethical concerns, a desire to protect something because it looks nice and enriches our lives, a need to maintain genetic diversity, the need to keep systems complex so they are more robust to environmental change, and economic incentives such as safari tourism or potential medicine or new food sources. Conservation management may be focused at the ecosystem, habitat or species level. Strategies can involve legislation to ban the hunting of a certain creature or even banning human entry into special reserves. Ensuring there are corridors for species to travel between patches within ecosystems can be another strategy for conservation.

Effort can be put into conserving the status quo, by reintroducing species or function back into an ecosystem to try to restore it. There may also be special strategies to try to ensure there is a large gene pool to protect against environmental change. Since many domesticated crops and animals would not survive in the wild there is concern that a disease or other disaster may come along that destroys those domesticated species and we would not have enough of the more robust wild ancestors left from which to develop the food of the future. This latter reason for conservation is clearly about the survival of humans and has an economic incentive.

Since conservation can be emotive there are moves to try to create rational measures by which effort and resources can be allocated. One method adopts the **ecosystem services** approach. Ecosystem services are those that support humans in some way. They may be:

- supporting services such as soil formation, photosynthesis, primary production, nutrient cycling and water cycling;
- provisioning services such as food, fibre, fuel, chemicals, medicines and fresh water;
- regulating services such as flood regulation, climate regulation, water purification, disease regulation, pollination and natural hazard regulation; and
- cultural services such as recreation, spiritual enrichment, learning, reflection and aesthetic values.

Evaluating the services that ecosystems provide to society in these ways can help focus attention on where investment and change might be required. If water is scarce and therefore valuable, and it is realised that, if the upstream ecosystem management was altered, then water would be secured, then this may aid management decisions. Investment in those changes may be worthwhile despite the fact that another service (e.g. provision of wood for furniture) might be reduced. The economics would allow a weighing-up of whether the furniture or the water was more valuable and payments might be forthcoming to change the management and perhaps even compensate the foresters or furniture makers, as this would still be cheaper than having to obtain water from a different location. Of course the problem with putting things into economic terms is that there is still some subjectivity in certain areas. For example it is difficult to put an economic value on the cultural services of archaeological preservation or the spiritual significance of a landscape. Therefore, there needs to be a broader, balanced view taken as to how to evaluate the service provision by different ecosystems to enhance their conservation. Nevertheless, raising awareness of the wider services that ecosystems provide appears to be of benefit and allows people to appreciate and value the services offered by quite distant systems and understand why those systems need protecting.

The **ecological footprint** is another tool for allowing people to understand the wider impact they have on the environment through the activities they undertake and the products they consume. It is similar to the carbon footprint discussed in Chapter 2. The ecological footprint estimates the amount of ecological resources that are used by individuals, companies or countries. It is a measure of the amount of biologically productive area needed to both produce resources used and to absorb waste created by an activity or in the creation of a purchased product. Most developed countries appear to be using more resources than they can sustain and are therefore running at an ecological deficit. It was estimated in 2010 by the Global Footprint Network that it takes the Earth 17 months to regenerate what we currently use in 12 months. The measure can be used to set goals for reducing consumption (recycling, wind power etc.) or increasing ecological productivity (for example, roof gardens).

SUMMARY

- Light, temperature, moisture availability, geological factors, humans and biotic factors such as competition, adaptation and migration all result in differences across the biosphere.
- Ecosystems are dynamic with inputs of energy, cycles of nutrients and changes through time.
- The cold biomes of the tundra and taiga have low productivity and vegetation is slow-growing restricted by short growing seasons. Irregular fire is a feature helping to regenerate and increase productivity in the taiga.
- The temperate biomes of the deciduous forest, evergreen forest, Mediterranean chaparral, and temperate grasslands are dominated by seasonality, with the latter two showing more adaptations to fire.
- The main biomes in the tropics are the highly productive and diverse tropical forests, the lower productivity savanna and the low productivity hot deserts. Savanna species often show adaptation to regular fire.
- Humans have had a major impact on the biosphere by reducing biodiversity through deforestation, agriculture, overexploitation of species and other environmental damage.

- Humans have assisted the movement of species around the world both intentionally and unintentionally often with adverse consequences and have increased the rate of evolution for crops and livestock.
- Climate change is likely to drive shifts in biomes and provide major challenges for conservation.
- Properly evaluating the full range of services that ecosystems provide to humans provides a way of evaluating ecosystem management strategies and increases the impetus to ensure ecosystems are sustainably managed.

FURTHER READING

Cox, C.B. and Moore, P.D. (2005) *Biogeography: An Ecological and Evolutionary Approach*, seventh edition, Chichester: Wiley-Blackwell. This is a widely used textbook.

Dickinson, G. and Murphy, K. (2007) *Ecosystems*, second edition, Abingdon: Routledge. An excellent textbook covering ecosystem concepts such as energy and material flows, ecosystem disturbance, succession and human impacts. There is also a very good section on biomes.

Ladle, R.J. and Whittaker, R.J. (eds) (2011) *Conservation Biogeography*, Oxford: Wiley-Blackwell. This text tackles the geographical approaches to conservation, mapping of ecological features and the design of conservation projects.

Wilson E.O. (1999) *The Diversity of Life*, second edition, New York: W.W. Norton and Company. This award winning book provides excellent bedtime reading and is also highly informative.

THE CHALLENGES AHEAD

Our study of physical geography tells us that the Earth is dynamic. Over long timescales the tectonic plates move. This has long-term impacts on the Earth's climate system, biogeography and landforms. However, it also affects the needs of society today, including protection from tectonic hazards such as earthquakes, volcanoes and tsunamis. We know that the climate will change. Study of the Earth's landforms shows us how they have been shaped by ice sheets, glaciers and periglacial activity many times in the past in areas where there is no ice today. Sea levels have risen and fallen in tandem with the retreat and advance of the ice. Even without human intervention there will be climate change. However, the changes brought about by humans to the Earth's atmospheric composition and its climate are exceptional. Equally the changes brought about by humans to soils, water pathways, river flow and channel change, water quality and coastal sediment dynamics are huge. It is hard to see how these changes will not be compounded in the near future as there are major challenges ahead. For example, we need to sort out the food and water supply for the world's growing population as it rises from 6 billion to 9 billion by around 2050. This needs to be done without devastating lots of ecosystems which might deliver fundamental services. We need to tackle climate change and its impacts through mitigation and adaptation. These challenges require international co-operation in ways we have not witnessed before.

Co-operation is needed because the Earth has a climate system which affects all nations. Technology, resources and trade mean we live in a global world, with a global economy. The need for co-operation was exemplified by a theory called the **tragedy of the commons** outlined by Garrett Hardin in the journal *Science* in 1968 based on ancient stories. This theory involves a field that anyone can use. Each farmer would be expected to keep as many cattle as possible on the field for their own gain. However, the logic of this common resource will bring tragedy. Each farmer seeks to maximise their gain and thinks about adding one more cow to the herd. However, this action has a negative and a positive effect. The positive effect is because the farmer earns money from that additional animal. However, the negative effect is the additional overgrazing created by one more animal. Because the effects of overgrazing are shared by all the farmers using the field, the negative effect for that farmer is therefore minor. The problem is that the farmer may think about adding more animals, coming to the same conclusion each time. Indeed, all of the farmers will think in a similar rational way. This leads to the tragedy. Each farmer exists within a system that encourages them to increase their herd without limit but in a field with limited resources. The field will be massively overgrazed and all of the vegetation will be removed. Therefore, none of the animals will survive and the farmers will be ruined.

This example may seem odd as you would think that the farmers would realise that there is a limited resource in the field and collectively they would try to sustainably share the resource provided by the field. Indeed this is often the case and collective or institutional management with careful rules or traditions which people are expected to follow can avoid overexploitation. But it is not such an odd example since it is quite analogous to allowing all people or countries the freedom to pump pollutants into the atmosphere or oceans when and where they want and in whatever quantity they want as if the atmosphere and oceans were unlimited resources. We have finite resources and we must therefore manage those resources. Communicating the nature of the problem to all those who may be affected is essential. This is where the role of physical geography comes in. The study of physical geography can be used to inform environmental managers and politicians of potential threats to the environment and potential solutions too. As such it can inform and shape political decision-making. Physical geography provides the scientific insight and understanding of the interconnected components of the Earth's system. It allows us to understand what resources we have and how they are affected by different processes occurring on different parts of the planet.

The environment is dynamic and we know from the material covered in this book that complex feedback mechanisms operate on Earth that sometimes throw up surprising rates and directions of change. A gradual slow change, which allows lots of time to think about a management solution, may suddenly become a rapid change if the system has reached a threshold and it suddenly jumps out of one stable state. This is a bit like a deltaic river that suddenly shifts its course due to a slow gradual build up of sediment clogging the channel which no longer provides an optimum route for flow. For another example in the Earth's system, look at the discussion on the thermohaline circulation system in Chapter 2. These examples tell us that the physical environment and environmental change is often uncertain. However, management action is needed early, probably well before there is a clear understanding of how things are going to change. This is known as the **precautionary principle**. There is risk involved in taking early action because it might physically be the wrong action and it may also be costly to implement and may not have been required. However, often the costs and physical consequences of not taking early action are enormous. The discussion on the economic impacts of climate change outlined by Sir Nicholas Stern as discussed towards the end of Chapter 2 is a good example.

Fossil fuel is the main energy source that we use today. Developing nations which account for most of the world's population are increasing fossil fuel use and this is likely to continue as they develop, even if the rich nations stabilise or reduce their emissions. Nuclear power carries safety concerns and is expensive but is a way to reduce greenhouse gas emissions. However, there are other ways forward. Nations have been meeting to discuss international agreements on emissions. Many countries are setting targets to reduce carbon use in energy provision. New energy sources are being sought and technology improved to harness wind, wave, biochemical and solar energy. Care is needed though to fully think through the impacts of what appear to be low carbon resources. The palm oil biofuel example discussed in Chapter 2 is a case that should raise concern. More broadly, taking up land for biofuels or other products, that might otherwise be used for agriculture may become a big issue when we are expecting to increase the human population by 50 per cent within 40 years. So international collaboration will be needed not only to deal with climate change but for the provision of food and provision of water; these are not just local issues, they are global issues which require urgent action. Physical geographers have an enormous amount to contribute to the present and the future in terms of understanding and dealing with these major challenges.

Research in physical geography has told us an enormous amount about how the world works. It can inform management and policymaking by providing insights into the integrated way Earth surface processes operate and the feedbacks that often exist. The understanding we have gained from research on the Earth's systems has led to the international meetings on climate change, action on new energy technologies, the quick international action in the 1980s and 1990s on CFCs to deal with the ozone hole, laws to tackle overfishing, protect soils, develop integrated land management solutions to floods, water quality, coastal management and so on. While we have been damaging the environment for quite some time, we have now taken notice and international effort is being put into tackling the issues. Just as the Earth's climate system, geomorphological processes and biosphere are interrelated, the world's major challenges around climate change, energy provision, poverty, malnutrition, disease, and hostile feelings around differences in the standard of living between countries are interrelated. Empowerment comes from understanding how human–environment relationships are interrelated and enables governments and people to take action thoughtfully, minimising negative feedback effects. This need to understand processes that operate close to the Earth's surface, in order to provide information for environmental management, should be major driver of research.

The focus of this short concluding section of the book has been about the applied nature of physical geography, the need for collaborative environmental management and the need for science to inform management action. However, I end on a note that says we also need serendipitous physical geography. I write as a research scientist (indeed as The Chair of Physical Geography at the University of Leeds) who is passionate about making sure that science has an impact on society. However, we must allow curiosity-driven research into environmental processes to thrive. Often many amazing discoveries are made that have huge impacts on society when there is a thriving research community able to follow new avenues and find out how the world works without there necessarily being a particular societal impact in mind. The inventors of the laser were just investigating some of Einstein's theories and did not foresee its practical use for eye surgery or DVD players. Microwave ovens and penicillin were outcomes of research on something completely different from their use today. So I encourage you to further study physical geography not just because you might *need* to know how components of the Earth's system work for your occupation or to pass an exam, but because you *want* to know how the Earth's system works. You might just make a serendipitous discovery that impacts your life and perhaps the lives of the rest of us too.

GLOSSARY

Ablation zone   The part of the glacier where there is a net loss of mass annually.

Active layer   The layer of the ground above permafrost that is seasonally frozen and thawed.

Aeolian transport   The movement of a substance by the wind.

Aerosols   Microscopic particles contained within the atmosphere that interfere with the Earth's incoming energy from the Sun and the outgoing energy being emitted from the Earth's surface. Aerosols can result in either cooling or warming of the Earth's climate.

Albedo   The proportion of incoming energy from the Sun that is reflected by a surface. Snow has a high albedo, whereas tarmac has a low albedo.

Aquifer   Rock that is porous enough to absorb and retain water and permeable enough to allow groundwater to pass through it freely.

Arête   A narrow mountain ridge between two adjacent cirques.

Atoll   An oceanic feature consisting of a central lagoon surrounded by a coral reef.

Baseflow   The stable portion of a river's discharge that is fed by groundwater.

Bases   Substances that have a pH higher than 7 and release hydroxide ions (OH$^-$).

Berm   The limit of the swash zone at the back of the beach, which exhibit a steep slope and a flat top.

Biodiversity   The number and variety of different plant and animal species within an ecosystem.

Biogeography   The study of spatial patterns and processes associated with life on Earth.

Biomass   The amount of soil and above-ground vegetation containing living organisms and plant roots.

Carbon footprint The amount of carbon that an individual or an organisation uses over a given time period. It is often expressed in terms of the equivalent amount of carbon dioxide.

Catchment area The area of land that drains to a given point on a river or lake.

Cation exchange capacity A measure of a soil's ability to hold and release certain elements, which is dependent on the net negative charge of the clay minerals within it.

Chemosynthesis The use of energy from oxidation of chemicals to manufacture carbohydrates for living organisms from water and carbon. The process uses chemical nutrients to produce energy for carbohydrate production rather than the Sun's energy as used in photosynthesis.

Cirques Basins carved into mountain sides by cirque glaciers. They possess steep rock walls, a rock basin and terminal moraine.

Cohesion The attraction between molecules in a substance that holds the substance together.

Col A hollow that forms in an arête from localised erosion, often providing a pass between mountain peaks.

Convection The molecular motion responsible for the transfer of energy, such as heat, through a fluid.

Convergent plate boundary Where two tectonic plates collide, creating major physical features, such as mountains, and hazards such as earthquakes and volcanoes.

Coriolis effect The rotation of the Earth on its axis causes a moving object or fluid to be apparently deflected. Deflection occurs to the right in the northern hemisphere and to the left in the southern hemisphere and is stronger at higher latitudes.

Corries See cirques.

Crag and tail features Formed as a glacier slides over a resistant rock mass ('crag') depositing sediment on the downstream side ('tail'), creating a streamlined feature.

Delta Large accumulation of sediment where a river enters the sea or a lake creating a raised river mouth. Deltas often have many branching channels that feed out from the main river.

Desertification The advancement of desert-like conditions due to human action or climate change resulting in soils with organic matter contents below 1.7 per cent.

Dew point The temperature at which a cooling air parcel becomes saturated with water vapour, at which water vapour condenses to form liquid water.

Divergent plate boundaries   Where two tectonic plates move apart and create new crust, leading to features such as the ocean floor, a mid-ocean ridge and rift valleys.

Drumlins   Streamlined mounds of till in the direction of ice flow, with a blunt upstream end and a tapered downstream end.

Dry adiabatic lapse rate   Air is a poor conductor of heat, therefore the cooling or warming of an air parcel is considered to be adiabatic (self-contained) as there is no exchange of energy between the air parcel and the surrounding atmosphere. The dry adiabatic lapse rate is 9.8°C per kilometre and applies only to air parcels that have not reached dew point, at which condensation occurs and the saturated adiabatic lapse rate is applied.

Ecological footprint   An estimation of the amount of ecological resources used by an individual, company or nation.

Ecological niche   Where there are no competitors for any of the resources required by an organism, it can occupy the ideal conditions to which it is adapted within that ecological community. No two organisms can occupy the same niche.

Ecosystem services   Services provided by ecosystems that support human life in some way, e.g. food, medicine and clean water. Evaluating these services helps to focus public attention on environmental issues that could result in the loss of these services.

El Niño Southern Oscillation   A reduction in upwelling of cold deep water in the South American Pacific resulting from reduced strength of trade winds over the equatorial Pacific Ocean, which in turn reduces the strength of westward-driven currents. This leads to unseasonal warm weather and disruption of pressure and precipitation systems in the southern hemisphere approximately every 5 years.

Environmental lapse rate   The normal rate of temperature change with altitude. Air temperature falls by approximately 6.4°C per kilometre, but is subject to variation.

Erratics   Large boulders that have been transported a long distance from their original source by glacial ice.

Eskers   A long, narrow, wavy ridge containing fine material transported by R-channels at the glacier bed. Eskers can be 20 to 30 metres high and up to 500 kilometres in length.

Evapotranspiration   The movement of water (liquid) from the Earth's surface into the atmosphere (as water vapour) by evaporation plus transpiration from plants (see definition below).

Fjord   A deep valley carved out by a glacier which has now become inundated by water from the sea.

Föhn wind   The warm, dry wind that blows down the lee side (downside) of a mountain, causing the lee side of mountain ranges to be significantly drier. The dryness occurs because water held by an air mass will be released as it rises, cools and condenses over mountains to form precipitation.

Food chain   The feeding links between species within an ecosystem indicating which species eat which other species.

Frost creep   The downslope movement of the active layer. The soil expands perpendicular to the sloping surface when it freezes, and settles vertically on thawing, causing an overall movement downslope.

Gas hydrates   Solid crystal structures in which the molecules of gas, such as methane, are combined with molecules of water. Their appearance resembles that of ice.

Gelifluction   Creep of melted ground which is saturated by meltwater above a permanently frozen layer. During the short seasonal thaw in periglacial regions the upper soil layer is inundated with a plentiful supply of water which helps cause the upper soil to slowly move.

Genetic modification   The scientific alteration of DNA, where genes are deleted or added in order to change certain characteristics of a species.

Geomorphology   The study of landforms and their characteristics so that their origin, development and history may be understood.

Geyser   A hot spring that periodically jets water into the air.

Glacials   Long, cold phases during the Quaternary (last 2.4 million years), which saw the widespread advancement of glaciers and ice sheets.

Gleying   The process by which a gley soil is produced. This happens when iron and manganese compounds within the soil are subject to stagnant, wet conditions and starved of oxygen so that the compounds are 'reduced'. The result is a blue-grey soil.

Gondwanaland   An ancient continent which was once made up of the joined-up land masses of present day Australia, Antarctica, South America and Africa.

Groyne   A long wooden, concrete or stone barrier typically constructed at right angles to a beach or river in order to trap sediment.

Gully   A small valley eroded sharply into soil by running water.

Gyre   A large, circular, rotating ocean current.

Hanging valleys   A side valley to a main glacial valley that terminates at a higher elevation than the floor of the main valley due to glacial erosion in the main valley.

Herbivores   An organism which consumes plant material only.

Holocene   The last 10,000 years to the present day, an interglacial (warm period) of the Quaternary.

Humus   Soil organic matter which is very resistant to decomposition.

Hydrograph   A graph showing the change in water discharge over time for a water body.

Ice shelves   A thick, floating mat of ice formed where ice flowing from the land reaches the ocean. The floating ice remains connected to the ice on land.

Ice wedges   A V-shaped wedge of ice that forms in areas with no insulating snow cover. When the ground becomes very cold it cracks to create frost crack polygons. The cracks become filled with water, which then freeze creating a wedge of ice which can grow through time, expanding the crack.

Igneous rock   Rock that has formed from melted material.

Infiltration capacity   The upper rate at which water can flow into the soil from the surface. This rate can change through time depending on how wet the soil is and the surface conditions.

Infiltration-excess overland flow   Where the rate of rainfall or irrigation water supply to the soil surface exceeds the infiltration capacity, leading excess water to flow across the surface. This is also known as Hortonian overland flow.

Infiltration rate   The time it takes for a unit of water added to a surface to enter a unit of soil.

Infrared radiation   The energy that is released by all solids, liquids and gases as heat.

Interglacials   Long, warm phases of the Quaternary (last 2.4 million years), in which glaciers and ice sheets receded and became limited to a few locations.

Intertropical convergence zone   The region where the trade winds from the northern and southern hemispheres converge. Conditions are favourable for warm, moist, rising air, resulting in cloudiness and heavy rainfall.

Jet streams   Narrow, high speed winds caused by sharp temperature gradients, located within Rossby waves in the upper atmosphere. They can be thousands of kilometres long and hundreds of kilometres wide.

Karst   A landscape shaped by the weathering of limestone rock, characterised by underground drainage tunnels and surface depressions.

Kettle holes   Depressions in the surface of a glacial sediment deposit caused by blocks of ice which become surrounded by sediment and, after melting, leave a hole in the sediment.

Keystone species   A species which is connected to most levels of the food chain, which if lost could result in collapse of the food web and major loss of biodiversity.

Lapse rate The rate at which air temperature decreases with increasing altitude.

Laterisation Warm weather and plentiful rainfall result in fast weathering conditions and leaching of material within soils. Laterite soils are produced which are often orange or red in colour. These soils often have little organic matter since decomposition and leaching removes this material quickly.

Leaching The removal of dissolved soil material vertically through the soil profile by surplus water.

Litter Soil organic matter consisting of decomposing plant and animal debris.

Longshore drift The transport of sediment along the coast by longshore currents which surge towards the beach at an oblique angle, followed by a backwash that transports sediment at right angles to the coast, resulting in a zigzag movement of material along the coast.

Macropore flow The transfer of water through the soil between large pores greater than 0.1 millimetres in diameter.

Marl Loose, earthy deposit with high concentrations of limestone.

Massive ice Isolated bands or lenses of ice several metres thick within soils.

Matrix flow The transfer of water through the soil between microscopic pores smaller than 0.1 millimetres in diameter.

Meander A bend in a river channel.

Metamorphic rock Rock which has been subject to heat and pressure which has resulted in a change to its structure or composition.

Mineralisation The release of plant nutrients during decomposition of organic matter, which can then be used by other organisms.

Moraines Linear mounds of glacial till that have been transported by a glacier. They are classified according to the method of their deposition.

Mud pot A small, hot spring consisting of bubbling mud.

Natural greenhouse effect Atmospheric greenhouse gases, such as carbon dioxide and water vapour, absorb 90 per cent of the long-wave radiation emitted from Earth, resulting in an average global temperature approximately 35°C warmer than would be experienced without the natural levels of greenhouse gases present.

Neap tides When the Sun opposes the gravitational pull of the Moon on the Earth (i.e. the Earth is positioned between the Sun and the Moon) the tidal range is reduced, resulting in lower high tides and higher low tides. It usually occurs during the first and third quarters of the Moon.

Occluded front   A faster, steeper cold front overtakes a slower warm front and lifts the mass of warm air upwards.

Organic farming   Farming which prohibits or minimises the use of human-made fertilisers and pesticides, and relies on natural biological processes to maintain soil fertility and control pests.

Oxbow lake   A crescent-shaped section of river channel cut off from the main river channel by sediment deposition and the slow migration of the channel location.

Peds   Naturally occurring clumps of soil particles.

Periglacial   Cold conditions with intense frost action but where there are no glaciers.

Permafrost   A zone in soil which has been frozen for more than two years.

Photosynthesis   The process by which organisms such as plants and algae (autotrophs) create carbohydrates and release oxygen using light energy, carbon dioxide and water.

Pingos   An ice-cored mound of earth that can reach 60 metres high and 500 metres in length, which are only found in periglacial areas. They are caused by the doming of the frozen ground beneath a former water body.

Planform   The shape of the river channel as viewed when looking down from the air.

Plant functional types   A way of classifying plant species based on their traits and association with specific environmental variables. Plant functional types are used by environmental models to predict how certain groups of species might respond to climate change.

Plate tectonics   A theory that concerns the large-scale movements of the Earth's crust (lithosphere). The crust is split into several 'plates', categorised as either oceanic or continental, which are moved by convection within the liquid mantle upon which they float. The movement of the plates is responsible for mountain building, trenches and earthquakes.

Ploughing boulders   Boulders found on periglacial slopes that move slowly downslope pushing through the soil, leaving a trough behind them and creating a bulge of sediment in front of the boulder. The movement is thought to occur due to the difference in thermal conditions beneath the boulder compared with its surroundings.

Podzolisation   Rainwater in cool climates flows through a thick upper litter layer to produce an organic rich leachate which then flows downwards through the soil collecting aluminium and iron. The soil becomes very acidic as a result.

Polar front The boundary at which cold air from the polar cell and warm tropical air from the Ferrel cell meet, causing air to rise.

Precautionary principle A decision-making approach which believes that lack of scientific evidence for warnings about future threats of serious damage should not be used as an excuse to avoid action in order to prevent damage from happening; action should be taken as early as possible.

Precipitation The condensation of water vapour to form water droplets in the atmosphere, which are then deposited on the Earth's surface in a liquid (e.g. rain) or solid form (e.g. snow, hail).

Pressure-melting point The melting point of solids such as ice is not constant, but varies according to pressure. The melting point of ice becomes lower with increasing pressure, meaning that under thicker ice there is likely to be water.

Protalus ramparts A linear arrangement of coarse sediment at the base of a periglacial slope, created by frost shattering of slope material which slides down over the snow pack and settles below it.

Quaternary The last 2.4 million years to the present day, characterised by the expansion and contraction of ice sheets in predictable cycles.

Reef (coral) Underwater landform constructed from the remains of corals and typically consisting of diverse ecosystems.

Refraction A process caused by a reduction in velocity as a wave enters shallow water, resulting in the wave front changing direction and 'bending' as it reaches the shore.

Regelation When ice meets an obstacle, such as a rock, pressure increases on the upstream side of the rock. This lowers the pressure melting point, resulting in melting of the ice on the upstream side of the rock, which then flows around the obstacle and refreezes on the downstream side due to the pressure melting point being higher. This allows ice to flow around obstacles.

Regolith The layer of soil overlying the bedrock which contains unconsolidated weathered parent material which provides the raw material for soil development.

Riffles The accumulation of coarse sediment which forms bar deposits across a river, which tend to be spaced between five and seven times the channel width apart.

River regime The variability of river flow over time, typically characterised over a year.

Roche mountonées Smaller versions of stoss-and-lee forms.

Rock glaciers A tongue-shaped body of rock and angular sediment that flows very slowly downslope like a glacier. Ice often occurs within pore spaces between the rock particles, which aids movement.

Rossby waves Large upper-atmosphere undulations which disturb the belt of prevailing westerly winds associated with the Ferrel cell. They contain jet streams.

Safety factor (also known as factor of safety) The ratio of the forces resisting movement to the forces promoting movement of material downslope. If the value is below 1, movement will occur.

Salinisation The collection of soluble salts within a soil that can have a detrimental impact on soil fertility. It concerns salts of sodium, magnesium and calcium.

Salinity The concentration of salt dissolved in water.

Saltating A method of sediment transport whereby sediment grains are bounced along a bed surface.

Saturated adiabatic lapse rate See *dry adiabatic lapse rate* for explanation of 'adiabatic'. The saturated adiabatic lapse rate is applied to air parcels that have reached dew point, which causes water vapour within the air parcel to condense into liquid water. This process releases heat and warms the air parcel, meaning the saturated adiabatic lapse rate is less than the dry adiabatic lapse rate. The saturated adiabatic lapse rate varies according to the temperature and moisture content of the air.

Saturation–excess overland flow Where all of the pore spaces within the soil become filled with water, and therefore saturated, forcing the excess water to flow across the surface.

Savanna A grassland environment, with a low density of trees, which is hot with an annual dry season.

Sedimentary rock Rock which had developed as sediment has accumulated over time and then become compressed and slowly formed solid matter.

Segregated ice A lens of ice found just below the active layer, which grows because of migration of water from around the lens to the freezing point.

Sesquioxides Contain three oxygen atoms and two atoms (radicals) from a different substance.

Shear stress A stress acting upon a particle in the same direction as the surface it is resting upon. In rivers the shear stress is the velocity of flowing water. When a sediment particle can be lifted from the river bed then the critical flow velocity (critical shear stress) is reached.

Shoaling A gradual decrease in water depth that causes waves to become higher as they move towards the land.

Shore platforms A flat or gently sloping wave-cut platform found at the base of a rocky cliff cut into the rock.

Soil horizons Layers of soil within a soil profile, distinctive in terms of colour and texture, created by the leaching and deposition of soil materials caused by water moving vertically through the profile.

Solifluction The slow movement of soil material downslope, where a saturated thawed layer moves across a layer of permafrost under gravity.

Solute flux The total movement of dissolved material through a system as measured by mass (e.g. kilograms).

Speciation The evolution of a new species, whereby small changes in the characteristics of successive generations leads to a species that is different from the ancestor from which it originated.

Spit A narrow strip of sand that protrudes into the sea, usually curved in a seaward direction, with one end attached to the mainland. Spits occur where the shore direction changes, e.g. at the mouth of an estuary.

Spring tides A tide that occurs on or around the time of a new or full Moon, which is characterised by unusually high or low tides. It occurs when the Sun and Moon are in alignment, reinforcing the gravitational pull of the Moon on the Earth.

Stack A steep column of rock protruding above the sea.

Stoss-and-lee forms Smoothed rock outcrops formed by basal sliding of a glacier, streamlined in the direction of glacier flow, with a plucked steep side facing downstream and a tapered end facing upstream.

Striations Small erosional features, with an appearance similar to scratches, caused by basal sliding of a glacier with embedded particles at the base of the ice, flowing over a rock surface and creating grooves.

Subduction zone Subduction can occur where two oceanic plates are forced together and one slides beneath the other, or where an oceanic plate is forced beneath the less dense continental plate. The plate material is then melted by the mantle beneath the Earth's crust.

Sublimation The process by which a solid changes directly into a gas.

Succession The changes to the structure and make-up of an ecological community over time.

Surf zone The area where the water depth becomes too shallow for waves, causing them to break.

Tarn A small lake that occupies the basin of a cirque.

Temperature inversion Air temperature increases with altitude, usually where a layer of warmer air overrides a layer of cooler air. The reverse of the normal environmental lapse rate.

Thermohaline circulation Large-scale deep ocean circulation involving vertical and lateral movements of large parcels of water, driven by gradients of water density which results from variations in water temperature and salinity.

Thermokarst A collection of irregularly spaced thaw lakes and depressions as a result of the melting of segregated ice.

Throughflow The movement of water draining through the soil in a downslope direction.

Tidal current The flow of water that is produced by the rise and fall of the tides, which is most pronounced in river mouths, estuaries and where flow is squeezed through inlets.

Trade winds Strong winds that flow from east to west towards the equator, between 30° north and south. They are deflected to the west due to the Coriolis effect.

Tragedy of the commons A theory that explains how communal resources can be degraded due to the selfish nature of individuals who use more than their fair share. It is used as an analogy for the unsustainable use of finite ecological resources.

Transpiration The evaporation of water through the pores of plant leaves which is released into the atmosphere.

Trophic levels Groupings of organisms within a food chain (see definition above). Primary producers (photosynthesising plants which form organic material from the Sun's energy, carbon dioxide and water) are at the lowest trophic level and these are fed upon by creatures at the second trophic level and so on.

Troposphere The lower layer of the atmosphere that extends between 6 and 15 kilometres in altitude above the Earth's surface.

Tsunami An energetic sea wave triggered by an earthquake, landslide or meteor impact in the ocean, which can become very large once it reaches shallow water, and cause devastation in coastal zones.

Water table The upper limit of the saturated zone of the soil or rock.

Watershed This term is often used in two ways. In many parts of the world it refers to the area of land that drains into one point in a river or lake. However, in other parts of the world the term can also mean the divide between an area that drains one way and an area that drains another.

Whaleback forms Smoothed rock outcrops formed by the sliding of a glacier, streamlined in the direction of glacier flow, with a smoothed steep side facing upstream and a tapered end facing downstream.

Yardangs Hills that have been smoothed and streamlined by erosion from dust carried by the wind.

INDEX

Page numbers in **bold** denote figures.